CW00407211

Sustainability in Agriculture

ISSUES IN ENVIRONMENTAL SCIENCE AND TECHNOLOGY

PREVIOUS TITLES IN THE SERIES:

1: Mining and its Environmental Impact
2: Waste Incineration and the Environment
3: Waste Treatment and Disposal
4: Volatile Organic Compounds in the Atmosphere
5: Agricultural Chemicals and the Environment
6: Chlorinated Organic Micropollutants
7: Contaminated Land and its Reclamation
8: Air Quality Management
9: Risk Assessment and Risk Management
10: Air Pollution and Health
11: Environmental Impact of Power Generation
12: Endocrine Disrupting Chemicals
13: Chemistry in the Marine Environment
14: Causes and Environmental Implications of Increased UV-B Radiation
15: Food Safety and Food Quality
16: Assessment and Reclamation of Contaminated Land
17: Global Environmental Change
18: Environmental and Health Impact of Solid Waste Management Activities
19: Sustainability and Environmental Impact of Renewable Energy Sources
20: Transport and the Environment

How to obtain future titles on publication

A subscription is available for this series. This will bring delivery of each new volume immediately on publication and also provide you with online access to each title *via* the Internet. For further information visit *www.rsc.org/issues* or write to:

Sales and Customer Care, Royal Society of Chemistry
Thomas Graham House, Science Park, Milton Road, Cambridge, CB4 0WF, UK
Telephone: +44(0) 1223 432360, Fax: +44(0) 1223 426017, E-mail: *sales@rsc.org*

ISSUES IN ENVIRONMENTAL SCIENCE AND TECHNOLOGY

EDITORS: R.E. HESTER AND R.M. HARRISON

21
Sustainability in Agriculture

RSC Publishing

ISBN 0-85404-201-6
ISSN 1350-7583

A catalogue record for this book is available from the British Library

Published by The Royal Society of Chemistry,
Thomas Graham House, Science Park, Milton Road,
Cambridge CB4 0WF, UK

Registered Charity Number 207890

For further information see our website at www.rsc.org

Typeset by Macmillan India Ltd
Printed by Biddles Ltd, Norfolk, UK

Preface

The World Trade Organisation (WTO) Ministerial Meeting which was held in Cancún, Mexico, in September 2003, was marked by mass demonstrations and protests against globalisation and the impacts of unfair trade on developing country agriculture. These public demonstrations received much attention in the world's newspapers and broadcast media. An estimated 10,000 activists marched in Cancún and a similar number of Mexican police were involved in securing the barricades separating activists from the WTO delegates. The self-immolation of Mr Lee Kyung Hae at the security barricade was widely reported – he was wearing a sign that declared 'the WTO kills farmers'.

The issues of 'free trade' and 'fair trade' are at the heart of the – often heated, sometimes violent – debate on world trade in agriculture. Massive subsidies, restrictive barriers, international collaboration and competition, and the livelihoods of millions of farmers around the world are involved. Concerns about sustainability in agriculture must inevitably take these issues into account, as well as changes in agricultural productivity and the impacts of novel developments such as genetically modified crops.

This volume aims to bring together many of the key issues which impact on agricultural sustainability in an endeavour to throw light on the subject and thereby promote informed and rational discussion of topics which so often generate powerful emotions and heated argument. A distinguished group of experts have contributed to the book from many different points of view and special interests. We believe this overall balanced assessment will contribute positively to the continuing debate.

The first article is by Jules Pretty who is professor in the Centre for Environment and Society at the University of Essex. Writing on recent progress and emergent challenges in agricultural practice, his review touches on food production and environmental costs, with a focus on pesticide use. This is followed in Chapter 2 by an overview of the ecological risks of transgenic plants by Paul Thompson, professor in the field of agriculture, food and community ethics at Michigan State University. Risk assessment and risk management in the context of the products of agricultural biotechnology and their ecological impacts are central themes of this article. The potential for health and socio-economic hazards implicit in the production of drugs and industrial chemicals from GM crops imposes a need for strict biological containment. These and other related concerns associated with the growth of transgenic crops are discussed in this chapter.

Nick Birch and Ron Wheatley, both members of the Entomology Group in the Scottish Crop Research Institute, take forward the discussion of transgenic plants in the next chapter on non-target impacts of GM pest-resistant crops. Their review of relevant scientific risk assessment studies focuses on the above- and below-ground interactions of *Bt* crops with their agro-ecosystems. Risks and benefits to non-target organisms, effects on important ecological functions and economic impacts are examined.

Chapter 4 presents a survey of land use change and sustainable development written by Dan Osborn from the Centre for Ecology and Hydrology at Lancaster University. This then is followed by a chapter by Ian Dickie of the Royal Society for the Protection of Birds and Anna Shiel of the National Trust which is focused on the UK environmental-economic consequences of decoupled European Union (EU) Common Agricultural Policy (CAP) payments. The June 2003 reform of the CAP introduced a phased single-farm payment for EU farmers, decoupled from production, to replace the old subsidy system and new requirements for basic agricultural and environmental standards were introduced. The environmental policy implications of this reform are examined with the aid of two specific case studies of its economic consequences.

The wider implications of agricultural subsidies are the subject of the final two chapters. In Chapter 6 James Smith of the University of Edinburgh's Centre of African Studies, has reviewed the impacts of unfair trade on developing country agriculture under the title 'Globalising Vulnerability'. Particular attention is given to a case study of sugar production in southern Africa countries. The book concludes with a chapter written by Colin Butler of the National Centre for Epidemiology and Population Health at the Australian National University. This discusses the moral and physical hazards associated with free trade in food. Although the general prognosis looks poor, a strong case is presented for a greater emphasis on fair trade and the need for greater consideration of moral issues in addition to economic ones in order to improve the lot of farmers and agricultural labourers in the less developed countries.

In summary, this volume of *Issues* presents an authoritative and balanced overview of many of the key factors that impact upon sustainability in agriculture. Its timeliness in treating hotly debated matters such as free trade, fair and unfair trade, GM crops, land use change and the economic consequences of recent changes in the CAP, make it essential reading for all those involved in agriculture. It will have particular value for farmers and students of farming, for policy makers, for environmental science students and teachers, and more broadly for all concerned about the future of agriculture worldwide.

Ronald E Hester
Roy M Harrison

Contents

Editors

Ronald E. Hester, BSc, DSc(London), PhD(Cornell), FRSC, CChem

Ronald E. Hester is now Emeritus Professor of Chemistry in the University of York. He was for short periods a research fellow in Cambridge and an assistant professor at Cornell before being appointed to a lectureship in chemistry in York in 1965. He was a full professor in York from 1983 to 2001. His more than 300 publications are mainly in the area of vibrational spectroscopy, latterly focusing on time-resolved studies of photoreaction intermediates and on biomolecular systems in solution. He is active in environmental chemistry and is a founder member and former chairman of the Environment Group of the Royal Society of Chemistry and editor of 'Industry and the Environment in Perspective' (RSC, 1983) and 'Understanding Our Environment' (RSC, 1986). As a member of the Council of the UK Science and Engineering Research Council and several of its subcommittees, panels and boards, he has been heavily involved in national science policy and administration. He was, from 1991 to 1993, a member of the UK Department of the Environment Advisory Committee on Hazardous Substances and from 1995 to 2000 was a member of the Publications and Information Board of the Royal Society of Chemistry.

Roy M. Harrison, BSc, PhD, DSc(Birmingham), FRSC, CChem, FRMetS, Hon MFPH, Hon FFOM

Roy M. Harrison is Queen Elizabeth II Birmingham Centenary Professor of Environmental Health in the University of Birmingham. He was previously Lecturer in Environmental Sciences at the University of Lancaster and Reader and Director of the Institute of Aerosol Science at the University of Essex. His more than 300 publications are mainly in the field of environmental chemistry, although his current work includes studies of human health impacts of atmospheric pollutants as well as research into the chemistry of pollution phenomena. He is a past Chairman of the Environment Group of the Royal Society of Chemistry for whom he has edited 'Pollution: Causes, Effects and Control' (RSC, 1983; Fourth Edition, 2001) and 'Understanding our Environment: An Introduction to Environmental Chemistry and Pollution' (RSC, Third Edition, 1999). He has a close interest in scientific and policy aspects of air pollution, having been Chairman of the

Department of Environment Quality of Urban Air Review Group and the DETR Atmospheric Particles Expert Group as well as a member of the Department of Health Committee on the Medical Effects of Air Pollutants. He is currently a member of the DEFRA Air Quality Expert Group, the DEFRA Advisory Committee on Hazardous Substances and the DEFRA Expert Panel on Air Quality Standards.

Contributors

A.N.E. Birch, *Scottish Crop Research Institute, Dundee, Scotland, DD2 5DA, UK*

Colin D. Butler, *National Centre for Epidemiology and Population Health, Australian National University, Canberra, Australia 0200*

Ian Dickie, *RSPB, The Lodge, Sandy, Bedfordshire, SG19 2DL, UK*

Daniel Osborn, *Centre for Ecology and Hydrology, Lancaster University, Bailrigg Campus, Lancaster, LA1 4AP, UK*

Jules Pretty, *Department of Biological Sciences, University of Essex, Colchester, Essex, CO4 3SQ, UK*

Anna Shiel, *The National Trust, 36 Queen Anne's Gate, London, SW1H 9AS, UK*

James Smith, *Centre of African Studies, The University of Edinburgh, 21 George Square, Edinburgh EH8 9LD, UK*

Paul B. Thompson, *Department of Philosophy, Michigan State University, East Lansing, MI 48824-1032, USA*

R.E. Wheatley, *Scottish Crop Research Institute, Dundee, Scotland, DD2 5DA, UK*

Sustainability in Agriculture: Recent Progress and Emergent Challenges

JULES PRETTY

1 Recent Progress on Food Production

There have been startling increases in food production across the world since the beginning of the 1960s. Since then, aggregate world food production has grown by 145%. In Africa, it rose by 140%, in Latin America by almost 200%, and in Asia by a remarkable 280%. The greatest increases have been in China – an extraordinary five-fold increase, mostly occurring in the 1980s and 1990s. In the industrialised regions, production started from a higher base – yet it still doubled in the USA over 40 years, and grew by 68% in western Europe.

Over the same period, world population has grown from three to six billion. Again, though, *per capita* agricultural production has outpaced population growth. For each person today, there is an extra 25% more food compared with people in 1960. These aggregate figures, however, hide important differences between regions. In Asia and Latin America, *per capita* food production increased by 76% and 28%, respectively. Africa, though, has fared badly, with food production per person 10% less today than in 1960. China performs best, with a trebling of food production per person over the same period. These agricultural production gains have lifted millions out of poverty and provided a platform for economic growth in many parts of the world.

However, these advances in aggregate productivity have not brought reductions in incidence of hunger for all. In the early 21st century, there are still some 800 million people hungry and lacking adequate access to food. A third are in East and South-East Asia, another third in South Asia, a quarter in Sub-Saharan Africa, and 5% each in Latin America/Caribbean and in North Africa/Near East. Nonetheless, there has been progress, as incidence of under-nourishment stood at 960 million in 1970, comprising a third of people in developing countries at the time. Since then, average *per capita* consumption of food has increased by 17% to 2,760 kilocalories per day–good as an

Issues in Environmental Science and Technology, No. 21
Sustainability in Agriculture

average, but still hiding a great many people surviving on less: 33 countries, mostly in Sub-Saharan Africa, still have *per capita* food consumption under 2,200 kcal/day.[1]

Despite great progress, things will probably get worse for many people before they get better. As total population continues to increase, until at least the mid 21st century, so the absolute demand for food will also increase. Increasing incomes will also mean people will have more purchasing power, and this will increase demand for food. But as diets change, so demand for the types of food will also shift radically. In particular, increasing urbanisation means people are more likely to adopt new diets, particularly consuming more meat and fewer traditional cereals and other foods, what has been called the nutrition transition.[2]

One of the most important changes in the world food system will come from an increase in consumption of livestock products. Meat demand is expected to rise rapidly, and this will change many farming systems. Livestock are important in mixed production systems, using foods and by-products that would not have been consumed by humans. But increasingly farmers are finding it easier to raise animals intensively, and feed them with cheap, though energetically-inefficient, cereals and oils. Currently, *per capita* annual food demand in industrialised countries is 550 kg of cereal and 78 kg of meat. By contrast, in developing countries it is only 260 kg of cereal and 30 kg of meat. These food consumption disparities between people in industrialised and developing countries are expected to persist.[3]

2 What is Agricultural Sustainability?

What do we understand by agricultural sustainability? Many different terms have come to be used to imply greater sustainability in some agricultural systems over prevailing ones (both pre-industrial and industrialised). These include sustainable, ecoagriculture, permaculture, organic, ecological, low-input, biodynamic, environmentally-sensitive, community-based, wise-use, farm-fresh and extensive. There is continuing and intense debate about whether agricultural systems using some of these terms qualify as sustainable.

Systems high in sustainability are making the best use of nature's goods and services whilst not damaging these assets.[4–11] The key principles are to:

i. integrate natural processes such as nutrient cycling, nitrogen fixation, soil regeneration and natural enemies of pests into food production processes;
ii. minimise the use of non-renewable inputs that damage the environment or harm the health of farmers and consumers;
iii. make productive use of the knowledge and skills of farmers, so improving their self-reliance and substituting human capital for costly inputs;
iv. make productive use of people's capacities to work together to solve common agricultural and natural resource problems, such as for pest, watershed, irrigation, forest and credit management.

The idea of agricultural sustainability does not mean ruling out any technologies or practices on ideological grounds. If a technology works to improve productivity for farmers, and does not harm the environment, then it is likely to be beneficial on sustainability grounds. Agricultural systems emphasising these principles are also multi-functional within landscapes and economies. They jointly produce food and other goods for farm families and markets, but also contribute to a range of valued public goods, such as clean water, wildlife, carbon sequestration in soils, flood protection, groundwater recharge and landscape amenity value.

As a more sustainable agriculture seeks to make the best use of nature's goods and services, so technologies and practices must be locally adapted and fitted into place. These are most likely to emerge from new configurations of social capital, comprising relations of trust embodied in new social organisations, new horizontal and vertical partnerships between institutions, and human capital comprising leadership, ingenuity, management skills and capacity to innovate. Agricultural systems with high levels of social and human assets are more able to innovate in the face of uncertainty.[12–13]

A common, though erroneous, assumption has been that agricultural sustainability approaches imply a net reduction in input use, and so are essentially extensive (they require more land to produce the same amount of food). All recent empirical evidence shows that successful agricultural sustainability initiatives and projects arise from changes in the factors of agricultural production (*e.g.* from the use of fertilisers to nitrogen-fixing legumes; from the use of pesticides to emphasis on natural enemies). However, these have also required reconfigurations on human capital (knowledge, management skills, labour) and social capital (capacity to work together).[14]

A better concept than extensive, therefore, is to suggest that sustainability implies intensification of resources – making better use of existing resources (*e.g.* land, water, biodiversity) and technologies. For many, the term intensification has come to imply something bad – leading, for example, in industrialised countries, to agricultural systems that impose significant environmental costs.[15–17] The critical question centres on the 'type of intensification'. Intensification using natural, social and human capital assets, combined with the use of best available technologies and inputs (best genotypes and best ecological management) that minimise or eliminate harm to the environment, can be termed 'sustainable intensification'.

3 The Environmental Challenge

Most commentators agree that food production will have to increase in the coming years, and that this will have to come from existing farmland. But solving the persistent hunger problem is not simply a matter of developing new agricultural technologies and practices. Most hungry consumers are poor, and so simply do not have the money to buy the food they need. Equally, poor producers cannot afford expensive technologies. They will have to find new types of solutions based on locally-available and/or cheap technologies combined with making the best of natural, social and human resources.

Increased food supply is a necessary though only partial condition for eliminating hunger and food poverty. What is important is who produces the food, has access to the technology and the knowledge to produce it, and has the purchasing power to acquire it. The conventional wisdom is that, in order to increase food supply, efforts should be redoubled to modernise agriculture. But the success of industrialised agriculture in recent decades has masked significant negative externalities, with environmental and health problems increasingly well-documented and costed, including Ecuador, China, Germany, the Philippines, the UK and the USA.[16–23] These environmental costs change our conclusions about which agricultural systems are the most efficient, and indicate that alternatives that reduce externalities should be sought.

There are surprisingly few data on the environmental and health costs imposed by agriculture on other sectors and interests. Agriculture can negatively affect the environment through overuse of natural resources as inputs or through their use as a sink for pollution. Such effects are called negative externalities because they are usually non-market effects and therefore their costs are not part of market prices. Negative externalities are one of the classic causes of market failure whereby the polluter does not pay the full costs of their actions, and therefore these costs are called external costs.[24]

Externalities in the agricultural sector have at least four features: i) their costs are often neglected; ii) they often occur with a time lag; iii) they often damage groups whose interests are not well represented in political or decision-making processes; and iv) the identity of the source of the externality is not always known. For example, farmers generally have few incentives to prevent pesticides escaping to water bodies, the atmosphere and to nearby nature as they transfer the full cost of cleaning up the environmental consequences to society at large. In the same way, pesticide manufacturers do not pay the full cost of all their products, as they do not suffer from any adverse side effects that may occur.

Partly as a result of lack of information, there is little agreement on the economic costs of externalities in agriculture. Some authors suggest that the current system of economic calculations grossly underestimates the current and future value of natural capital.[25–26] Such valuation of ecosystem services remains controversial because of methodological and measurement problems[22, 27–29] and because of its role in influencing public opinions and policy decisions. The great success of industrialised agriculture in recent decades has masked significant negative externalities, many of which arise from pesticide overuse and misuse.

There are also growing concerns that such systems may not reduce food poverty. Poor farmers, at least whilst they remain poor, need low-cost and readily available technologies and practices to increase local food production. At the same time, land and water degradation is increasingly posing a threat to food security and the livelihoods of rural people who occupy degradation-prone lands. Some of the most significant environmental and health problems centre on the use of pesticides in agricultural systems.[30]

4 How Much Pesticide is Used?

In the past 50 years, the use of pesticides in agriculture has increased dramatically, and now amounts to some 2.56 billion kg per year. The highest growth rates for the world market, some 12% per year, occurred in the 1960s. These later fell back to 2% during the 1980s, and reached only 0.6% per year during the 1990s. In the early 21st century, the annual value of the global market was US $25 billion, down from a high of more than $30 billion in the late 1990s. Some $3 billion of sales are in developing countries.[31] Herbicides account for 49% of sales, insecticides 25%, fungicides 22%, and others about 3% (Table 1).

Table 1 *World and US use of pesticide active ingredients (average for 1998–99)*[48,49]

Pesticide Use	*World pesticide use*		*US pesticide use*	
	(Million kg of active ingredient)	*%*	*(Million kg of active ingredient)*	*%*
Herbicides	948	37	246	44
Insecticides	643	25	52	9
Fungicides	251	10	37	7
Other[a,b]	721	28	219	40
Total	2563	100	554	100

Notes

[a] Other includes nematicides, fumigants, rodenticides, molluscicides, aquatic and fish/bird pesticides, and other chemicals used as pesticides (*e.g.* sulfur, petroleum products)

[b] Other in the US includes 150 million kg of chemicals used as pesticides (sulfur, petroleum products)

A third of the world market by value is in the USA, which represents 22% of active ingredient use. In the US, though, large amounts of pesticide are used in the home/garden (17% by value) and in industrial, commercial and government settings (13% by value). By active ingredient, US agriculture used 324 million kg per year (which is 75% of all reported pesticide use, as this does not include sulfur and petroleum products). Use in agriculture has increased from 166 million kg in the 1960s, peaked at 376 million kg in 1981, and since fallen back. However, expenditure has grown. Farmers spent some $8 billion on pesticides in the USA in 1998–99, about 4% of total farm expenditures.

Industrialised countries accounted for 70% of the total market in the late 1990s, but sales are now growing in developing countries (Figure 1). Japan is the most intensive user per area of cultivated land. The global use of all pesticide products is highly concentrated on a few major crops, with some 85% by sales applied to fruit and vegetables (25%), rice (11%), maize (11%), wheat and barley (11%), cotton (10%), and soybean (8%).[30]

There is also considerable variation from country to country in the kinds of pesticide used. Herbicides dominate the North America and European domestic markets, but insecticides are more commonly used elsewhere in the world. In the USA in the late 1990s, 14 of the top 25 pesticides used were herbicides (by kg active ingredient (a.i.)), with the most commonly used products being atrazine (33–36 million kg), glyphosate (30–33 million kg), metam sodium (a fumigant, 27–29 million kg), acetochlor (14–16 million kg), methyl bromide (13–15 million kg), 2,4-D (13–15 million kg), malathion (13–15 million kg), metolachlor (12–14 million kg), and trifluran (8–10 million kg). Glyphosate and 2,4-D were the most common products used in domestic and industrial settings (EPA, 2001). In Asia, 40% of pesticides are used on rice, and in India and Pakistan some 60% are used on cotton. India and China are the largest pesticide consumers in Asia. Pesticide consumption in Africa is low on a per hectare basis.

5 The Benefits of Integrated Pest Management (IPM)

Recent IPM programmes, particularly in developing countries, are beginning to show how pesticide use can be reduced and pest management practices can be modified

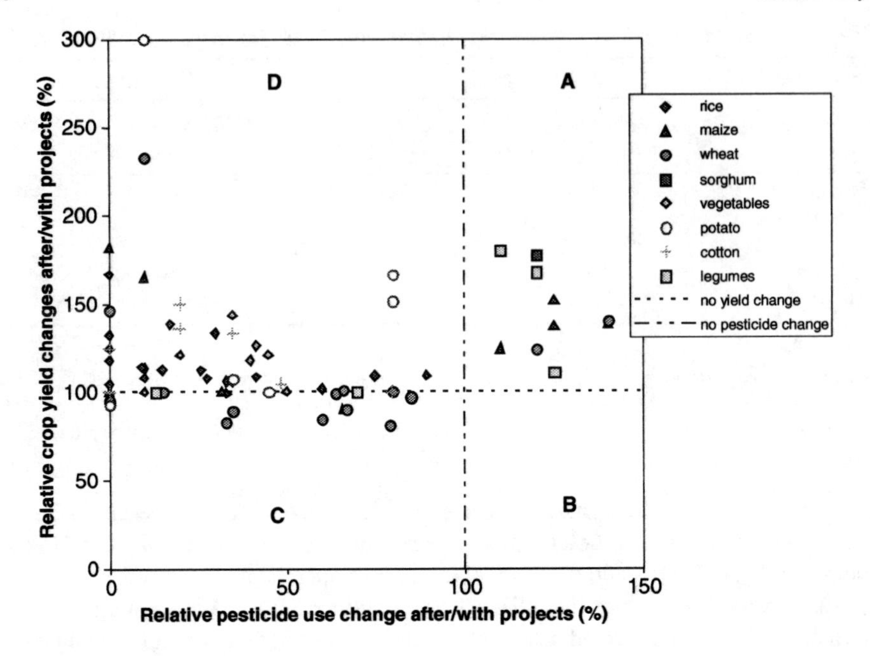

Figure 1 *Association between pesticide use and crop yields (data from 81 studies of crops, 62 projects, 26 countries)*[33]

without yield penalties. In principle, there are four possible trajectories of impact if IPM is introduced:

i. both pesticide use and yields increase (A);
ii. pesticide use increases but yields decline (B);
iii. both pesticide use and yields fall (C);
iv. pesticide use declines, but yields increase (D).

The assumption of conventional agriculture is that pesticide use and yields are positively correlated. For IPM, the trajectory moving into sector A is therefore unlikely but not impossible, for example in low input systems. What is expected is a move into sector C. While a change into sector B would be against economic rationale, farmers are unlikely to adopt IPM if their profits would be lowered. A shift into sector D would indicate that current pesticide use has negative yield effects or that the amount saved from pesticides is reallocated to other yield increasing inputs. This could be possible with excessive use of herbicides or when pesticides cause outbreaks of secondary pests, such as observed with the brown plant hopper in rice.[32]

Figures 1 and 2 show data from 62 IPM initiatives in 26 developing and industrialised countries (Australia, Bangladesh, China, Cuba, Ecuador, Egypt, Germany, Honduras, India, Indonesia, Japan, Kenya, Laos, Nepal, Netherlands, Pakistan, Philippines, Senegal, Sri Lanka, Switzerland, Tanzania, Thailand, UK, USA, Vietnam and Zimbabwe). Pretty and Waibel[33] used an existing dataset that audits progress

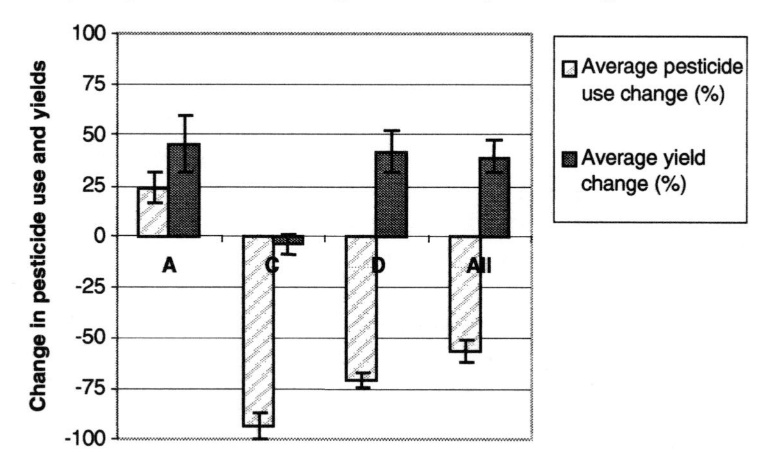

Figure 2 *Changes in pesticide use and yields in 62 projects (A: n = 10; C: n = 5; D: n = 47)*[33]

being made on yields and input use with agricultural sustainability approaches. The research audited progress in developing countries, and assessed the extent to which farmers were increasing food production by using low-cost and locally available technologies and inputs.

The 62 IPM initiatives cover some 25.3 million ha, *i.e.* less than 1% of the world crop area, and directly involve some 5.4 million farm households. The evidence on pesticide use is derived from data on both the number of sprays per hectare and the amount of active ingredient used per hectare. This analysis does not include recent evidence on the effect of genetically modified crops, some of which have resulted in reductions in the use of pesticides, such as herbicides in the UK[34] and China,[35] and some of which have led to increases, such as in the USA.[36]

There is only one sector B case reported in recent literature.[37] Such a case has recently been reported from Java for farmers who received training under the popular UN Food and Agriculture Organisation's Farmer Field School model. However, the paper does not offer any plausible explanation for this result but does point out that there were administrative problems in implementing the project which was funded by the World Bank. The cases in sector C, where yields fall slightly while pesticide use falls dramatically, are mainly cereal farming systems in Europe, where yields typically fall up to some 80% of current levels when pesticide use is reduced to 10–90% of current levels.[38]

Sector A contains ten projects where total pesticide use has indeed increased in the course of IPM introduction. These are mainly in zero-tillage and conservation agriculture systems, where reduced tillage creates substantial benefits for soil health and reduced off-site pollution and flooding costs. These systems usually require increased use of herbicides for weed control,[39] though there are some examples of organic zero-tillage systems in Latin America.[40] Over 60% of the projects fall into category D where pesticide use declines and yields increase. While pesticide reduction is to be expected, as farmers substitute pesticides with information, yield increases induced by IPM is a more complex issue. It is likely, for example, that farmers who receive good quality field training will not only improve their pest management skills but also become more efficient in other agronomic practices such as water, soil and nutrient management. They can also invest some of the cash saved from pesticides in other inputs such as higher quality seeds and inorganic fertilisers.

The baseline against which change is measured has been current practices, which as we have shown are known to cause some harm to environments and human health. A change that reduces specifically the use of those pesticides causing harm (*e.g.* WHO Class I and II products) can create large benefits for farmers and other groups in society. In the light of these results, the question must now be: why, if IPM shows clear benefits to farmers why are reduced pesticide use approaches and technologies not more widely in use? Or, as Wilson and Tisdell[41] have put it: why are farmers still using so many pesticides? While, as cited above, these authors have provided some possible explanation, some questions remain in the light of successful country cases where pesticide reductions have actually worked. In the next section we summarise these cases.

6 Current Evidence of Pesticide Reductions at Country Level

A growing number of countries are now reporting reductions in pesticide use as a result of the adoption of agricultural sustainability principles. These have occurred as a result of two very different types of approach:

i. policy-led and primarily top-down pesticide-reduction programmes in industrialised countries, such as in Sweden, Denmark, the Netherlands and some provinces in Canada;
ii. farmer field school-led and policy-supported community IPM in rice programmes, beginning in South East Asia, and then spreading throughout Asia and then to other continents.

Several OECD countries set ambitious national targets in the mid-1990s to reduce the use of inputs. Sweden's aim was to reduce input consumption by 20% by the year 2000. The Netherlands also sought a cut in pesticide use of 50% by the year 2000 as part of its Multi-Year Plan for Crop Protection. The cost of this reduction programme was estimated at $1.3 billion, most of which was raised by levies on sales. Denmark aimed for a 50% cut in its pesticide use by 1997, a plan which relied mostly on advice, research and training. Canada set a target for a 50% reduction in pesticide use by 2000 in Quebec and by 2002 in Ontario. In the USA, the administration announced in 1993 a programme to reduce pesticide use whilst promoting sustainable agriculture. The aim was to see some form of IPM on 75% of the total area of farmland by the year 2000.

Supplemented by other policy measures, such as new regulations, training programmes, provision of alternative control measures and reduced price support, there have been some considerable reductions in input use. In Sweden, pesticide consumption fell by 61% between 1981–85 and 1996–2000 (from 23 to 9.3 million kg a.i.); in Denmark, by 40% between 1985 and 1995 (from 7 to 4.3 million kg a.i.); in the Netherlands, by 41% between 1985 and 1995 (from 21.3 to 12.6 million kg a.i.), and in Ontario by 40% between 1983 and 1998 (from 8.8 to 5.2 million kg). However, the full significance of these apparent sharp falls in use is disputed. In Sweden, half the decline was attributed to the introduction of new lower dose products, such as the use of sulfonurea products applied only at 0.004–0.006 kg a.i. ha^{-1} instead of phenoxyherbicides applied at 1–2 kg a.i. ha^{-1}. In Denmark, reduction was not accompanied by a cut in the frequency of application, which remains at the 1981 level of 2.5 doses ha^{-1} yr^{-1}. Success has been achieved without a diminished dependence on pesticides, which should really embody the basic concept for pesticide reduction.

Another analysis of Swedish pesticide consumption used LD_{50} values as an indicator of acute toxicity, and this showed that the changes had resulted in a fall from 38,000 acute toxicity equivalents to 11,000 and then to 8,700 by the end of the 1990s.[42] However, another measure, the hectare-dose method (the quantity of active ingredient applied per ha) showed a substantial increase – from 1.6 million in 1981–85 to 4.3 million in 1996–2000.

Scientific studies in industrialised countries underline the findings from pesticide reduction programmes that pesticide reduction can be beneficial to society's goals. For example, both private and public benefits were generated from IPM adoption for early leaf spot on peanuts in Virginia.[43] The total savings in external costs were estimated to be $844,000 per year for 59,000 households, on top of which farmers benefited from a small but important reduction in inputs costs. It was calculated that the 40% reduction in pesticide use between 1983 and 1998 in Ontario produced benefits of CAN $305 per household.[44] Aggregated across all 3.78 million households in the province, the value of the environmental risk reduction was $1.18 billion.

While IPM has had mixed success in the industrialised world it has received much more attention in the developing world. Here the discovery that pest attack on rice was proportional to the amount of pesticides used had a significant impact.[32] They found that pesticides were killing the natural enemies (spiders, beetles, parasitoids) of insect pests, and when these are eliminated from agro-ecosystems, pests are able to expand in numbers rapidly. This led in 1986 to the banning by the Indonesian government of 57 types of pesticides for use on rice, combined with the launching of a national system of farmer field schools to help farmers learn about the benefits of biodiversity in fields. The outcomes in terms of human and social development have been remarkable, and farmer field schools are now being deployed in many parts of the world.

In Bangladesh, for example, a combined aquaculture and IPM programme is being implemented by CARE with the support of the UK government and the European Union. Six thousand farmer field schools have been completed, with 150,000 farmers adopting more sustainable rice production on about 50,000 hectares.[45] The programmes also emphasise fish cultivation in paddy fields, and vegetable cultivation on rice field dykes. Rice yields have improved by about 5–7%, and costs of production have fallen owing to reduced pesticide use. Each hectare of paddy, though, yields up to 750 kg of fish, a substantial increase in total system productivity for poor farmers with very few resources. Similar effects are seen with rice aquaculture in China.[8]

Such substantial changes in pesticide use are bringing countries economic benefits in the form of avoided costs. One of the first studies to quantify the social costs of pesticide use was conducted at the International Rice Research Institute (IRRI) in the Philippines. Researchers investigated the health status of Filipino rice farmers exposed to pesticides, and found statistically significant increased eye, skin, lung and neurological disorders. Two-thirds of farmers suffered from severe irritation of the conjunctivae, and about half had eczema, nail pitting and various respiratory problems.[18, 46] In addition, the authors showed that in a normal year no insecticide application was better than researcher-recommended economic thresholds or even farmers' routine practices.

A so-called 'complete protection' strategy, with nine pesticide sprays per season, was not economical in any case. When health costs were factored in, insecticide use in rice became completely uneconomical. As Rola and Pingali[46] put it: 'the value of crops lost to pests is invariably lower than the cost of treating pesticide-related illness and the associated loss in farmer productivity. When health costs are factored in, the natural control option is the most profitable pest management strategy'. Any expected positive production benefits of applying pesticides are clearly overwhelmed by the health costs.

Other studies have calculated the economic benefits to farmers and wider society of IPM and pesticide reduction programmes. IPM in onion production in the Philippines reduced pesticide use by 25–65% without reducing yields.[47] Farmers benefited through increased incomes, and it was estimated that some $150,000 worth of benefits were created for the 4,600 residents of the five villages within the programme area.

Through various multi-lateral agreements, most countries in the world have indicated they are in favour of the idea of agricultural sustainability. Clearly, there are now opportunities to extend policy-led programmes including farmer field schools for pesticide reduction and to increase farmers' knowledge of alternative pest management options across diverse agricultural systems.

7 The Wider Policy Context for Agricultural Sustainability

Three things are now clear from evidence on the recent spread of agricultural sustainability:

i. some technologies and social processes for local scale adoption of more sustainable agricultural practices are increasingly well-tested and established;
ii. the social and institutional conditions for spread are less well-understood, but have been established in several contexts, leading to more rapid spread in the 1990s and early 2000s;
iii. the political conditions for the emergence of supportive policies are least well-established, with only a very few examples of real progress.

As indicated earlier, agricultural sustainability can contribute to increased food production, as well as make an impact on rural people's welfare and livelihoods. Clearly much can be done with existing resources. A transition towards a more sustainable agriculture will not, however, happen without some external help and money. There are always transition costs in learning, in developing new or adapting old technologies, in learning to work together, and in breaking free from existing patterns of thought and practice. It also costs time and money to rebuild depleted natural and social capital.

Most agricultural sustainability improvements seen in the 1990s and early 2000s arose despite existing national and institutional policies, rather than because of them. Although almost every country would now say it supports the idea of agricultural sustainability, the evidence points towards only patchy reforms. Nonetheless, recent years have seen some progress towards the recognition of the need for policies to support sustainable agriculture. Yet only three countries have given explicit national support for sustainable agriculture – putting it at the centre of agricultural development policy and integrating policies accordingly.

These are Cuba, Switzerland and Bhutan. Cuba has a national policy for alternative agriculture; Switzerland has three tiers of support to encourage environmental services from agriculture and rural development; and Bhutan has a national environmental policy coordinated across all sectors.[1]

Several countries have given sub-regional support to agricultural sustainability, such as the states of Santa Caterina, Paraná and Rio Grande do Sul in southern Brazil supporting zero-tillage, catchment management and rural agribusiness development, and some states in India supporting participatory watershed and irrigation management. A larger number of countries have reformed parts of agricultural policies, such as China's support for integrated ecological demonstration villages, Kenya's catchment approach to soil conservation, Indonesia's ban on pesticides and programme for farmer field schools, Bolivia's regional integration of agricultural and rural policies, Sweden's support for organic agriculture, Burkina Faso's land policy, and Sri Lanka and the Philippines' stipulation that water users' groups be formed to manage irrigation systems.

A good example of a carefully designed and integrated programme comes from China. In March 1994, the government published a White Paper to set out its plan for implementation of Agenda 21, and put forward ecological farming, known as *Shengtai Nongye* or agro-ecological engineering, as the approach to achieve sustainability in agriculture. Pilot projects have been established in 2000 townships and villages spread across 150 counties. Policy for these 'eco-counties' is organised through a cross-ministry partnership, which uses a variety of incentives to encourage adoption of diverse production systems to replace monocultures. These include subsidies and loans, technical assistance, tax exemptions and deductions, security of land tenure, marketing services and linkages to research organisations. These eco-counties contain some 12 million hectares of land, about half of which is cropland, and though only covering a relatively small part of China's total agricultural land, they do illustrate what is possible when policy is appropriately coordinated.

An even larger number of countries have seen some progress on agricultural sustainability at project and programme level. However, progress on the ground still remains largely despite, rather than because of, explicit policy support. No agriculture minister is likely to say they are against sustainable agriculture, yet good words remain to be translated into comprehensive policy reforms. Agricultural systems can be economically, environmentally and socially sustainable, and contribute positively to local livelihoods. But without appropriate policy support, they are likely to remain at best localised in extent, and at worst simply wither away.

8 Areas of Debate and Disagreement

What we do not yet know is whether progress towards more sustainable agricultural systems will result in enough food to meet the current food needs in developing countries, let alone the future needs after continued population growth and adoption of more urban and meat-rich diets. But what is occurring should be cause for cautious optimism, particularly as evidence indicates that productivity can grow over time if natural, social and human assets are accumulated.

A more sustainable agriculture which improves the asset base can lead to rural livelihood improvements. People can be better off, have more food, be better organised, have access to external services and power structures and have more choices in their lives.

But like all major changes, such transitions can also provoke secondary problems. For example, building a road near a forest can help farmers reach food markets, but also aid illegal timber extraction. Projects may be making considerable progress on reducing soil erosion and increasing water conservation through adoption of zero-tillage, but still continue to rely on applications of herbicides. If land has to be closed off to grazing for rehabilitation, then people with no other source of feed may have to sell their livestock; and if cropping intensity increases or new land is taken into cultivation, then the burden of increased workloads may fall particularly on women. Also additional incomes arising from sales of produce may go directly to men in households, who are less likely than women to invest in children and the household as a whole.

New winners and losers will emerge with the widespread adoption of sustainable agriculture. Producers of current agrochemical products are likely to suffer market losses from a more limited role for their products. The increase in assets that could come from sustainable livelihoods based on sustainable agriculture may simply increase the incentives for more powerful interests to take over. Not all political interests will be content to see poor farmers and families organise into more powerful social networks and alliances.

Many countries also have national policies that now advocate export-led agricultural development. Access to international markets is clearly important for poorer countries, and successful competition for market share can be a very significant source of foreign exchange. However, this approach has some drawbacks:

i. poor countries are in competition with each other for market share, and so there is likely to be a downward pressure on prices, which reduces returns over time unless productivity continues to increase;
ii. markets for agri-food products are fickle, and can be rapidly undermined by alternative products or threats (*e.g.* avian bird flu and the collapse of the Thai poultry sector);
iii. distant markets are less sensitive to the potential negative externalities of agricultural production and are rarely pro-poor (with the exception of fair-trade products);
iv. smallholders have many difficulties in accessing international markets and market information.

There is indeed very little clear evidence that export-led poverty alleviation has worked. Even Vietnam, which has earned considerable foreign exchange from agricultural development, has had to do so at very low prices and little value added.

More importantly, an export-led approach can seem to ignore the in-country opportunities for agricultural development focused on local and regional markets. Agricultural policies with both sustainability and poverty-reduction aims should adopt a multi-track approach that emphasises five components: i) small farmer development linked to local markets; ii) agri-business development – both small businesses and export-led; iii) agro-processing and value-added activities – to ensure that returns are maximised in-country; iv) urban agriculture – as many urban people rely on small-scale urban food production that rarely appears in national statistics; and v) livestock development – to meet local increases in demand for meat (predicted to increase as economies become richer).

A differentiated approach for agricultural policies will become increasingly necessary if agricultural systems themselves are to become more productive whilst reducing negative impacts on the environment.

References

1. J. Pretty, *Agri-Culture: Reconnecting People, Land and Nature,* Earthscan, London, 2002.
2. B. Popkin, *Public Health Nutrition*, 1998, **1(1)**, 5–21.
3. C. Delgado, M. Rosegrant, H. Steinfield, S. Ehui and C. Courbois, *Livestock to 2020: The Next Food Revolution,* International Food Policy Research Institute, Washington DC, 1999.
4. M.A. Altieri, *Agroecology: The Science of Sustainable Agriculture*, Westview Press, Boulder, 1995.
5. J. Pretty, *Regenerating Agriculture: Policies and Practice for Sustainability and Self-Reliance*, Earthscan, London and National Academy Press, Washington, 1995.
6. G.R. Conway, *The Doubly Green Revolution*, Penguin, London, 1997.
7. National Research Council, *Our Common Journey*, National Academy Press, Washington DC, 2000.
8. L. Wenhua, *Agro-Ecological Farming Systems in China*, UNESCO, Paris, 2001.
9. J.A. McNeely and S.J. Scherr, *Ecoagriculture*, Island Press, Washington DC, 2003.
10. N.G. Röling and M.A. Ewagemakers (eds), *Facilitating Sustainable Agriculture*, Cambridge University Press, Cambridge, 1997.
11. N. Uphoff (ed), *Agroecological Innovations*, Earthscan, London, 2002.
12. N. Uphoff in P. Dasgupta and I. Serageldin (eds), *Social Capital: A Multiperspective Approach*, World Bank, Washington DC, 1998.
13. J. Pretty and H. Ward, *World Development,* 2001, **29 (2)**, 209–227.
14. J. Pretty, *Science,* 2003, **302**, 1912–1915.
15. G.R. Conway and J. Pretty, *Unwelcome Harvest: Agriculture and Pollution*, Earthscan, London, 1991.
16. J. Pretty, C. Brett, D. Gee, R. Hine, C.F. Mason, J.I.L Morison, H. Raven, M. Rayment and G. van der Bijl, *Agricultural Systems*, 2000, **65 (2)**, 113–136.
17. E. M. Tegtmeier and M.D. Duffy, *International journal of Agricultural Sustainability*, 2004, **2** (in press).
18. P.L. Pingali and P.A. Roger, *Impact of Pesticides on Farmers' Health and the Rice Environment*, Kluwer, Dordrecht, 1995.

19. C.C. Crissman, J.M. Antle and S.M. Capalbo (eds.), *Economic, Environmental and Health Tradeoffs in Agriculture*, CIP, Lima and Kluwer, Boston, 1998.
20. H. Waibel, G. Fleischer and H. Becker, *Agrarwirtschaft,* 1999, **48 H, 6, S,** 219–230.
21. J. Pretty, C. Brett, D. Gee, R.E. Hine, C.F. Mason, J.I.L. Morison, M. Rayment, G. van der Bijl and T. Dobbs. *Journal of Environment planning and Management* 2001, **44(2)**, 263–283.
22. J.Pretty, C.F. Mason, D.B. Nedwell and R.E. Hine, *Environmental Science and Technology,* 2003, **37(2)**, 201–208.
23. D. Norse, L.Ji, J. Leshan and Z.Zheng, *Environmental Costs of Rice Production in China*, Aileen Press, Bethesda, 2001.
24. W.J. Baumol and W.E. Oates, *The Theory of Environmental Policy*, Cambridge University Press, Cambridge, 1988.
25. R. Costanza, R. d'Arge, R. de Groot, S. Farber, M. Grasso, B. Hannon, K. Limburg, S. Naeem, R.V. O'Neil, J.Paruelo, R.G. Raskin, P.Sutton and M.van den Belt, *Nature,* 1997, **387**, 253–260.
26. G.Daily (ed), *Nature's Services: Societal Dependence on Natural Ecosystems*, Island Press, Washington DC, 1997.
27. S.Georgiou, I.H. Langford, I.J. Bateman and R.K. Turner, *Environment and Planning,* 1998, **30(4)**, 577–594
28. N. Hanley, D. MacMillan, R.E. Wright, C. Bullock, I. Simpson, D. Parrison and R. Crabtree, *Journal of Agricultural Economics,* 1998, **49 (1)** 1–15.
29. R.T. Carson, *Environmental Science and Technology,* 2000, **34**, 1413–1418.
30. J. Pretty (ed), *The Pesticide Detox*, Earthscan, London, 2004.
31. CropLife International, *Acutely Toxic Pesticides: Risk Assessment, Risk Management and Risk Reduction in Developing Countries and Economies in Transition* www.croplife.org, 2002.
32. P.E. Kenmore, F.O. Carino, C.A. Perez, V.A. Dyck and A.P. Gutierrez, *Journal of Plant Protection Tropics,* 1984, **1(1)**, 19–37.
33. J. Pretty and H. Waibel in J. Pretty (ed), *The Pesticide Detox,* Earthscan, London, 2004.
34. G.T. Champion, M.J. May, S. Bennett, D.R. Brooks, S.J. Clark, R.E. Daniels, L.G. Firbank, A.J. Haughton, C. Hawes, M.S. Heard, J.N. Perry, Z. Randle, M.J. Rossall, P. Rothery, M.P. Skellern, R.J. Scott, G.R. Squire and M.R. Thomas, *Philosophical Transactions of the Royal Society London,* 2003, **358**, 1801–1818.
35. Nuffield Council on Bioethics, *The Use of Genetically Modified Crops in Developing Countries*, London, 2004.
36. C.M. Benbrook, *Impacts of Genetically Engineered Crops on Pesticide Use in the United States*, Northwest Science and Environmental Policy Center, Ames, Iowa, 2003.
37. G. Feder, R. Murgai and J.B. Quizon, *Review of Agricultural Economics,* 2004, **26(1)**, 45–62.
39. H. de Freitas, in F. Hinchcliffe, J. Thompson, J. Pretty, I. Guijt and P. Shah. (eds), *Fertile Ground* IT Publications, London, 1998.
40. P. Petersen, J.M. Tardin and F. Marochi, *Environmental Development and Sustainability,* 2000, **1**, 235–252.
41. C. Wilson and C. Tisdell, *Ecological Economics,* 2001, **39**, 449–462.
42. G. Ekström and P. Bergkvist, *Pesticides News,* 2001, **54,** 10–11.

43. J.D. Mullen, G.W. Norton and D.W. Reaves, *Journal of Agriculture and Applied Economics* 1994, **29**, 243–253.
44. C. Brethour and A. Weerskink, *Agricultural Economics,* 2001, **25**, 219–226.
46. A. Rola and P. Pingali, *Pesticides, Rice Productivity, and Farmers' Health,* IRRI, Los Baños, Philippines, 1993.
45. M. Barzman and S. Desilles, in N. Uphoff (ed), *Agroecological Innovations,* Earthscan, London, 2002.
47. L.C.M. Cuyno, G.W. Norton and A. Rola, *Agricultural Economics,* 2001, **25**, 227–233.
48. OECD, *Environmental Outlook for the Chemicals Industry,* Paris, 2001.
49. EPA, *Pesticide industry sales and usage, 1998 and 1999 market estimates*, Environmental Protection Agency, Washington, DC, 2001.

Ecological Risks of Transgenic Plants: A Framework for Assessment and Conceptual Issues

PAUL B. THOMPSON

1 Introduction

Agricultural technology of all kinds is used in producing food and fibre, and the primary rationale for developing new agricultural technology derives from the obvious benefit that human beings derive from the reliable availability of food and fibre commodity goods. Over the centuries, new tools and farming methods have affected humanity's access to food and fibre goods in innumerable ways, though it is difficult to disaggregate impacts associated with technical improvements in transport and public health from those associated with agricultural technology proper. Yields from basic food crops have increased, leading to more reliable food supply and lower prices for consumers. Technical strategies for avoiding catastrophic crop losses have been developed. Agricultural technologies have also reduced the amount of human labour needed to produce crops and animals, or have made farm work safer and less onerous. Frequently, new agricultural technology has been developed to ameliorate residual problems created by a previous generation of technology that had been adopted because it was thought, on balance, to improve food and fibre production.

It can be exceedingly difficult to reach consensus on the net social value of changes in agricultural technology, even when such changes are in the past, their effects are generally known and there is substantial agreement on the facts. As the 2002 report of the National Research Council *Environmental Effects of Transgenic Crops* notes,

> 'Some U.S. citizens see the last 50 years of the twentieth century as a time when hundreds of years of insecurity over food availability came to an end. In their eyes, innovative technologies such as plant breeding, water management,

Issues in Environmental Science and Technology, No. 21
Sustainability in Agriculture

fertilizers, and synthetic pesticides played a heroic role in this drama. Others look back on the same events and see an era when for the first time in history human activity threatened the basic stability of global ecosystems on which all life, including human society, depends. In their eyes, modern agricultural science and technology are inimical to the natural environment.'[1]

Unlike some of the technologies mentioned in the NRC report, the development of recombinant DNA techniques to introduce nucleotide sequences into the genome of agricultural plants (*i.e. genetic engineering* or *genetic modification*) has been accompanied by significant debate over the wisdom of adopting such plants and animals for agricultural use. A significant component of this debate focuses on the likely consequences that production and consumption of such will have.

This chapter provides an introductory overview of concepts and terminology for reviewing the ecological risks associated with transgenic crops, as well as a general discussion of how risks and benefits might figure in evaluating transgenic crops. *Transgenic crops* are those developed using recombinant DNA methods for introducing novel traits. The broader term *agricultural biotechnology* will be used to indicate the new generation of reproductive technologies that are based upon the discovery and characterisation of recombinant DNA. These include methods for sequencing and manipulating DNA, especially the use of *agrobacter tumiferens* and ballistic methods of introducing nucleotide sequences into plant genomes. Plants developed using recombinant methods for introducing nucleotide sequences will be described as transgenic without regard to the source of these sequences. The term *transgene* will be used to refer to the nucleotide sequences so introduced without regard to whether these sequences code for proteins or perform regulatory functions, or indeed whether they are functional within the transgenic organism in any way.

Ecological risks are understood to include potentially adverse or harmful events that might occur by virtue of the presence or introduction of a given substance into a given environment. This includes harm *to* ecosystems as well as harm to human beings or human interests that occurs as a result of events transpiring through ecosystem processes. Exact specification of the harms or mechanisms relevant to ecological risk is difficult, potentially contentious and cannot be accomplished by a simple definition. Thus, further characterisation of ecological risk will be discussed below. However, ecological risk is typically interpreted to exclude hazards to human health brought about by intentionally consuming a substance as a food or drug, as well as adverse social or economic outcomes brought about by the operation of market forces. Food safety of transgenic crops and farmer profitability risks are excluded from the following discussion, except insofar as they are the result of inadvertent transport of transgenes outside the established food system. The established food system comprises intentional production, sale, transport and processing of transgenic seed and commodity grain produced from transgenic seed.

The aim of this chapter is to characterise a framework for conceptualising, measuring and evaluating ecological risks from transgenic crops. It does not discuss empirical findings relevant to biologically or socially based risks and benefits from transgenic crops, much less with respect to the likelihood that either beneficial or harmful outcomes will materialise in any specific case.

2 The Products of Agricultural Biotechnology

Although there has been extensive research and development of both microbial and animal applications of recombinant DNA, the focus will here be limited to plant biotechnology. So far, both public and private research organisations have developed, patented and in many cases marketed a large number of laboratory and crop development techniques using recombinant DNA, including methods for introducing and controlling gene products within plant systems, as well as a number of nucleotide sequences that may be introduced into future agricultural crops. Two of the most widely discussed transgenic crops, infertile (or so-called 'terminator') seeds and Vitamin A enhanced (or 'golden') rice have not been and may never be released for use by farmers. Transgenic crops currently being grown for commercial purposes include virus resistant varieties of squash and papaya, as well as a few crops (including maize) that have been developed to produce products for use outside the food chain. However, the majority of transgenic crops currently in production have been developed for resistance to the chemical herbicides glyphosate (*e.g. Round Up*) or glufosinate, or incorporate versions of *bacillus thuringiensus (Bt)* genes, which produce toxins that control infestations of caterpillars (but are not toxic to other species).

Herbicide tolerant varieties have been released by both public and private laboratories in a number of agricultural crops, including soy, cotton, canola and flax. *Bt* has been studied in a number of crops, but commercial applications currently in production are almost exclusively in maize and cotton, and are offered for sale by for-profit companies. *Bt* maize is currently estimated to make up approximately 40% of US maize production. Transgenic varieties of maize have also been developed for herbicide tolerance and for production of biologics (biologically active substances used by industry) and drugs. Although these varieties are in current production, acreage is small they have not been released for adoption by farmers. The pharmaceutical and industrial substances produced in maize crops are not approved for use in food, and the production of these crops is tightly controlled by the regulatory agencies.

In the future, transgenic crops may be developed for a large variety of purposes, including enhanced nutrition (such as golden rice), enhanced flavour or cooking characteristics, altered flowering control, disease resistance, and tolerance to soil and climate variations. However, many of the products expected to be placed in commercial use during the next decade will not be intended for use as human food or animal feed. These include more pharmaceutical or biologic producing crops like those discussed above, as well as crop varieties (including maize varieties) developed primarily for conversion to fuels. Some, but not all, of these non-food crops will need to be carefully segregated from the human food system. Many of these crops developed for pharmaceutical or industrial products will not be released as commercial varieties available to farmers, but fuel crops will have a large acreage.[1]

3 Using Risk Analysis to Evaluate Transgenic Crops

Over the past 25 years, the multi-disciplinary field of risk analysis has evolved to improve scientific methods for anticipating and managing unintended and unwanted outcomes of many kinds. A fairly standard approach to risk assessment has emerged within this field. Assessment of ecological risks can be usefully framed as a three-stage process consisting of *hazard identification, risk measurement* (or risk quantification), and *risk management*. It will also be important to identify processes of *risk communication* that should be conducted throughout the entire process of risk assessment.[2] Although the main elements of risk assessment are widely accepted, there are variations in the stages of this approach that reflect the particular problem to be studied, variations that reflect the use or application to which a risk assessment will be put, and also variations in terminology that can be a source of confusion when risk assessment results are introduced in public discussions. Each of these opportunities for alternative interpretation or variation of method can introduce ambiguity into the conceptualisation of ecological risk. Ambiguities of this kind have arguably plagued the characterisation of ecological risk from transgenic crops in both the scientific literature and the public debate. As such, a more detailed review of risk assessment with careful attention to alternative opportunities for interpreting its stages and terminology is appropriate.

Hazard identification includes a characterisation of the forms of danger, harm or injury that may be associated with the agent or activity in question, as well as a characterisation of the features thought to have the potential to cause danger, harm or injury. With such a characterisation in hand, it is possible to use a variety of scientific techniques to determine how likely it is that danger, harm or injury will actually materialise. Studies intended to assign or describe the probability that unwanted events or harms will actually occur represent the *risk quantification* phase of risk analysis. In describing these two activities as distinct phases of risk assessment, risk analysts make a key distinction between *hazard* or the potential for danger or of a harmful or injurious outcome, and *exposure,* or the analytic methods used to determine the probability that the hazard will actually materialise. *Risk* is subsequently said to be a function of hazard and exposure. Thus, being ill with a cold is a hazard of the winter season, but this hazard materialises only when events such as being in the presence of a rhino virus occur in conjunction with vulnerabilities in the immune system creating exposure to this hazard. The risk of a cold reflects both the degree and seriousness of the hazard, as well as the likelihood that the hazard will materialise. One key point to note is that simply because one has identified a hazard–a possible adverse or unwanted outcome–one has not necessarily identified a risk. Risk implies an estimate of both hazard and exposure.[2]

This terminology is applied in contexts that include the epidemiological investigation of disease outbreaks, the toxicological analysis of chemically or biologically active compounds and engineering assessments of failure modes for complex systems, such as aircraft flight controls or nuclear power plants. The actual activity carried out in hazard identification, risk quantification and risk management varies considerably in these contexts. In toxicological studies, hazard identification may involve experimental studies intended to characterise chemical or biological activity, while the assessment of exposure involves the estimation of dose-response relationships. In engineering risk assessment, hazard identification may involve a largely conceptual exercise of listing potential events regarded as adverse, while exposure is understood to be the sequence and combination of intervening events that must transpire in order for that potential to be realised. In both cases, the process of attempting to measure the probability of harmful or injurious impacts often gives rise to the recognition of new hazards, unanticipated in the early stages of risk assessment. As a result, there is typically a process of iteration between the stages of hazard identification and risk quantification.

Clearly, both hazard identification and risk quantification are subject to scientific uncertainty. Models or data for estimating the probability of an event may be incomplete. Even when this is not the case, the accuracy of predictions is subject to statistically measurable margins of error, and there can be differences of opinion about how such margins of error should be reflected in quantifying risk. Additionally, there remain large areas of ignorance in any attempt to predict outcomes, areas where science may simply not be capable of conceptualising, much less anticipating, an entire class of possible events. Differing views about how to respond to such uncertainties can lead to different approaches and differences of opinion about how to link the activities of hazard identification and risk quantification.

For example, within engineering contexts, the unpredictability of system performance under specified conditions might be recognized during the process of modelling system performance for exposure assessment. The inability to confidently predict how an aircraft control system will perform under given conditions will typically be defined as a hazard, and the exposure will be assessed by estimating how likely such conditions are to occur. In contrast, lack of reasonable certainty about the chemical or biological activity of substance is not typically regarded as in itself hazardous in a toxicological risk assessment, where recognition that there are always 'unknown unknowns' is considered to reflect the human condition. The difference here reflects both the respective degree of confidence that engineers and toxicologists give to their models and the value placed on predictable system performance in the context of assessing the risks associated with aircraft control systems. Extrapolation from relatively well-established domains of risk assessment to ecological risk requires careful evaluation of contextual elements that influence attitudes toward uncertainty and predictability of system performance.

Value judgments can be implicit within scientific characterisations of hazard and exposure. For example, there are cases where an outcome that is advantageous to one person or group is disadvantageous for others; hence there may be no neutral or objective way to characterise such outcomes as either beneficial or harmful. There are also straightforward disagreements about what should be counted as helpful or hurtful. This may, for example, be the case with respect to the simple occurrence of transgene migration. Is a transgene 'in the wrong place,' (*i.e.* anywhere beyond the crop into which it was deliberately introduced) already a harmful event, or does harm occur only when that transgene is maintained in a population, or has demonstrable adverse impact on human or non-target species? This is not the sort of question that the biological sciences are equipped to answer. Yet any attempt to use science in anticipating the consequences of technology demands some sort of provisional stance with respect to which outcomes are worth predicting. As such value judgments are always implicit even within the most neutral or scientifically objective attempts to characterise benefit and risk.[3,4]

Risk management is the process of deciding what to do about risks. The fact that a risk exists does not necessarily imply that anything should be done about it. Some risks are simply tolerated or accepted as a component of daily life. If there is a decision to take active steps in response to risk, a number of options are available.

Government agencies may take direct regulatory action to reduce the likelihood of harm or damage, or they may use strategies that empower private actors to do this. Policies that help industry or farmer groups to cooperate in risk-reducing activities (such as voluntary bans on the sale or use of a given substance) or that provide reliable information through standardised product labels are examples of the latter strategy. In some contexts it is possible to manage risk through insurance schemes that compensate those who experience damages. Many factors influence risk management, including the distribution of risk and benefit, the ability of risk bearers to obtain information and voluntarily accept or reject risk, as well as societal and cultural values that make certain types of risks (*e.g.* catastrophic or irreversible risks) seem more worthy of active management than others.

It is also important to place risks into a comparative context, so that one can be confident that steps to mitigate one risk do not create even greater risks from another source.[5] For example, minimising risk from agricultural biotechnology could lead to greater use of agricultural chemicals. If this were the case, a comparison of the relative ecological risks of these two options is needed in risk management. A recent meta-analysis of multiple studies concluded that total pesticide use reductions have occurred for US production of cotton and soybeans as a result of adopting transgenic varieties, though results for maize are more equivocal.[6] It is too early to know whether there may be additional unforeseen benefits that may arise in conjunction with first generation products from agricultural biotechnology. Such benefits are by their very nature unlikely to become evident until after technologies have been in use for some time. One example of a possible unforeseen benefit from *Bt* maize is speculation that *Bt* maize varieties may be more resistant to aflatoxin infestations,[7] though at present the reasons why this might be the case are unclear.

Both measured and speculative environmental benefits from transgenic crops highlight the fact that risk management may also involve weighing the risks associated with the release of any given transgenic crop against the benefits of release.[8] Some applications of risk assessment may add an explicit phase of *benefits assessment.* Like risk, benefits have two dimensions, the added or positive value associated with a possible outcome or state of affairs, and the probability that this value will actually materialise if a given course of action is taken. As the concept of risk implies a value judgment, so does the concept of benefit. As such, benefits assessment, like risk assessment, cannot be construed as a purely scientific activity. Methods for estimating economic benefits associated with agricultural production technologies have been used for many years, and can be applied to the assessment of farm profitability as well as to returns to food and fibre consumers.[9] These methods have been developed independently from risk assessment, and are frequently represented within an alternative framework for considering net social impact in which benefits are compared to costs. Methods for estimating environmental or social benefits are considerably less well developed. While risk management frequently does weigh benefit against risk, it is fair to say that the formal integration of benefits assessment into the risk assessment framework is less standard than the other four components.

Risk communication refers to a number of interactions that risk assessors and risk managers undertake throughout the process of analysing and managing risks. Experience has shown that public policies for managing risk often go awry because of poor communication with the broader public. There are several junctures in the process of identifying hazards, quantifying risks and developing risk management strategies where public input may be valuable. For example, it will be important to be sure that the assessment process does not neglect hazards that may be very obvious to affected parties, but unknown to technical experts who may have little knowledge of local circumstances. Such detailed knowledge of local farming methods or market structures will also be critical in the process of quantifying risks. Finally, risk assessment processes that appear to take place entirely behind closed doors can provoke anger and mistrust. Risk management strategies that neglect the role of public involvement can backfire, creating a greater perception of risk than ever before.[10]

The stages of risk assessment are sometimes represented in a sequential fashion, so that hazard identification and risk measurement are described as initial, scientific stages in the process that provide information to be used in risk management (Figure 1). While this representation provides insight into the underlying logic of risk assessment, it can be misleading, especially if 'risk communication' is added as a fourth box emerging from 'risk management.' In fact, risk management decisions can be undertaken with no formal processes of hazard identification or risk measurement, especially when uncertainties or the costs of acquiring such information are judged to be excessive. Furthermore, risk management processes penetrate into hazard identification and risk measurement, especially with regard to establishing priorities for which outcomes and exposure pathways are important to investigate. Placing 'risk communication' at the end of this process can convey the impression that communication is a one-way process whereby experts inform the public about risks. In fact, obtaining information from practitioners and affected parties is often a critical component of the other three phases of risk assessment. It is only in a few well-structured regulatory agencies operating under clear legal mandates stipulating which risks are actionable that risk assessment can be characterised as sequential activity of research into hazards and exposure followed by management decision-making and finally by dissemination of information. In fact, each phase of risk assessment is typically conducted in an iterative dialog with the other three. In situations where there is little agreement on priorities and problem identification, the distinction between one phase of risk assessment and another has less to do with the temporal sequence for conducting these activities than with the different tasks and methods appropriate to each. With these qualifications in mind, the diagram provides some insight into the underlying logic of risk assessment and its role in decision-making.

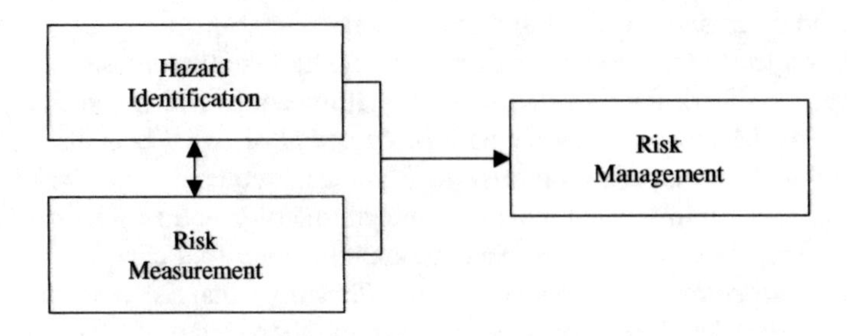

Figure 1 *The stages of risk assessment*

4 Ecological Hazards and Today's Transgenic Crops

In the present context, an ecological hazard is an unintended event regarded as adverse to the environment, as distinct from harm to human health or from socio-economic impact. Some possible impacts on the ecosystem from transgenic crops are not adverse. As described above, if adoption of a transgenic crop allowed for a decrease in the net application of toxic chemicals for pest protection, this impact would not (all things being equal) be regarded as adverse. Other possible impacts might be adverse but would not be regarded as unintended. A transgenic crop that extends the temperature range at which a plant such as the soybean can be planted will very likely displace other types of land use that might be environmentally preferable. Such intended impacts from agricultural production are not typically characterised as environmental risks.

For convenience, ecological hazards can be classified into two fairly comprehensive categories: *loss of or reduced ecosystem functioning* and *decreased biodiversity*, including genetic diversity. Loss of or reduced ecosystem functioning refers to effects on key ecological processes such as soil, water and nutrient cycles. Affects on microbiota or on the mix and complexity of organisms that alter soil formation would be one example of such an impact. Dramatic changes in water use might be associated with any number of possible transgenic modifications, and might affect rates of salinisation or the availability of water for other organisms in the environment. The category of decreased biodiversity includes a wide array of possible affects on non-target organisms that register as changes in the number and composition of organisms within an ecosystem. Such effects include toxicity to beneficial insects, loss or contamination of food and water supplies to wildlife, and displacement of either flora or fauna as a result of invasive properties that might be associated with a transgenic crop or with a wild relative affected by transgene migration. In addition, any decline in the genetic diversity within land races or within wild relatives would be regarded as an adverse effect on biodiversity. These effects may be direct or indirect. A direct effect may be an environmental toxicity from the transgene whereas indirect effects may be that the introduced trait leads the farmer to destroy more forest or consume more water.[11–13]

For completeness, it is important to reiterate that there is also the potential for offsetting environmental benefits that correlate with each of these categories of environmental risk. Thus, just as a transgenic crop poses a hazard with respect to biodiversity, it may also create the possibility of an increase in biodiversity if, for example, the introduction of the transgenic crop leads to a substantial reduction in the amount of land planted under very intensive production systems that provide relatively few opportunities for non-crop organisms to thrive. One measured impact of the herbicide-tolerant crops is an increase of low-till and no-till farming that increases soil organic matter and in-field and off-field biodiversity.

Before any of these hazards can be conceptualised as a risk, it is necessary to stipulate and then measure the likelihood associated with any sequence of events leading to the realisation of a hazard. Empirical studies are needed to produce a more detailed discussion, but it is useful to indicate three pathways that such a sequence of events would follow. The first of these begins with intentional release of the transgenic organism. What would follow from this is the introduction of transgene products into the environment at a rate that should be fairly straightforward to calculate. This provides the basis for anticipating phenomena such as toxicity and bioaccumulation that would be expected as a direct but unintended result of the introduction of gene products into the environment. A second possible route for exposure is associated with invasive, volunteer transgenic organisms. Invasives include crop plants originally introduced by farmers but capable of reproducing and spreading without cultivation, and also plants that are introduced into environments inadvertently through shipments or other movement of seed and grain. Invasive volunteers might lead to forms of toxicity and bioaccumulation, much like intentional introductions. They might also lead to unintended displacement of a species or community and subsequent effects on biodiversity. The third exposure scenario involves introgression of transgenes into feral populations or wild relatives. These feral populations might then become invasive, leading to displacement of a species or community, exposure to a hazard if there is toxicity, or decreases in genetic diversity.[13]

Exposure pathways for indirect effects are exceedingly complex. For example, it is possible that introduction of transgenic crops could be an important element in a scenario leading to greater use of chemicals, habitat conversion and agricultural expansion. Here, the environmental impacts would only indirectly be associated with a specific transgene, yet the net environmental affects from such scenarios might be much more significant than those associated with more direct impact. Quantification of such risks is a difficult and daunting task.

5 Ecological Exposure Pathways for Health and Socio-economic Hazards

In addition to adverse effects on ecosystems, ecological risk may be understood to include exposure to health or socio-economic hazards that occurs through ecological mechanisms. With respect to health, the most obvious example is the potential for environmental spread of transgenes known to involve human health hazards. A 2004 report from the Union of Concerned Scientists notes.[14]

> 'The production of drugs and industrial chemicals in corn and other food crops presents obvious risks. If genes find their way from pharm crops to ordinary corn, they or their products could wind up in drug-laced corn flakes. In addition, crops that unintentionally contain drugs or plastics could also prove harmful to domestic animals that eat contaminated feed; to deer, mice, birds and other wildlife that feed in pharm crop fields; or to organisms in the living soil.'

Strictly speaking, the concerns noted represent hazards, rather than risks, and hazards to human health or domestic animals would not normally be included in a review of ecological risks. It may be appropriate to do so, however, when the exposure to these hazards involves ecological mechanisms such as gene flow, invasion or other ecological means of transport for hazardous transgenes. Since the expectation is that the pharmaceutical and industrial crops noted in the report will be grown under strict biological containment, exposure pathways involving ecological mechanisms will also include events involving the breach or failure of these containment systems.

Events precipitating a breach of containment may arise from mechanical, biological or other system failure modes, but human factors must not be overlooked in assessing the likelihood of ecological dispersion of harmful transgenes. Contamination of the food supply with unapproved *Cry9C Bt* toxin (*e.g. Starlink*) occurred through predictable human failures in maintaining separation of animal feeds from the human food supply.[15] In 2001, two University of California professors published a scientific paper alleging the introgression of *Bt* transgenes into Mexican landraces of maize.[16] Although the paper became the subject of an extended controversy in its own right, transgenes not approved for use in Mexico could clearly have entered the fields of Mexican maize farmers who obtained specimens being sold legally for use as animal feed. Given this initial exposure event, ecological mechanisms could be involved in establishing the transgene in landraces.[17] Although in the former case the initial exposure event appears not to have interacted with ecological mechanisms to cause a persistence or spread of the *Bt* transgene, the impact in the latter case is currently unknown.

While the *Cry9C* event appears not to have eventuated in harm to human health, there is little dispute that it resulted in significant economic losses.[15] Other types of economic loss could occur if introgression of transgenes into crops reduces their value or saleability on international markets.[14] Less tangible forms of social hazard are associated with the feeling of loss that individual farmers and food consumers might feel if widespread ecological dispersion of transgenes forecloses their ability to express religious, cultural and personal values through the preparation and consumption of food. This loss is sometimes articulated as the loss of a 'right to choose'. Such hazards may be better understood as challenges to the legitimacy of social relations than as a quantifiable loss of value.[18] It is also possible that traditional rural communities experience a pervasive and fundamental kind of cultural identity in the form of specific farming practices, harvest and market patterns, and longstanding farm-to-table relationships.[19] If so, it is conceivable that unconstrained ecological spread of transgenes could threaten such forms of community solidarity. Although characterising such hazards under the rubric of ecological risk is likely to be controversial, people in almost all societies tend to articulate and conceptualise challenges to personal values and ways of life in terms of danger to health and well-being.[20]

Furthermore, when people feel that their values and concerns have been subverted in a systematic way, there is the potential for fairly widespread damage to public confidence in public and private institutions. Some of the public fears of genetic engineering are linked to fears, uncertainties or value judgments associated with the growth of powerful multi-national companies, increasing privatisation of intellectual property, and the integration of global trade and financial markets. Uncontrolled ecological spread of transgenes is not infrequently mentioned along with such hazards, and may be seen as closely connected with them by members of the general public. As such, there may be an interweaving of ecological risk scenarios with broad socio-political concerns that has pervasive impact on confidence in science and in public institutions. Several authors have examined the possibility that controversies over genetically engineered food might be having this kind of complex socio-cultural impact.[21,22] Such broad-ranging types of damage to social institutions are almost never addressed in formal risk assessment processes, and are not typically considered by authors reviewing ecological risk. Given the apparent significance of ecological concerns as a component of broad social attitudes, it may be appropriate to consider such socio-cultural hazards as events that could be indirectly triggered by transgene movement through ecological mechanisms. Inclusion of such socio-cultural hazards would provide a rationale for viewing a 'transgene in the wrong place' as an event precipitating an exposure sequence, even when no further ecological effect can be identified.

6 Strategies in Ecological Risk Management

The strategies that have been most widely used in public health contexts and for coping with environmental risk are instances of *risk optimisation*. A second, but long-standing and traditional approach to risk management has been to create social structures in which persons exposed to risk have opportunity to give or withhold consent to risk exposures. Although this approach has long been used in workplace and commercial settings, it has more recently been formalised for research contexts and can be characterised as *informed consent.* A third approach has emerged especially in debates over genetic engineering in agriculture calling for use of the *precautionary principle.*

The basic idea behind risk optimisation is that risk management should aim for an optimal (or at least satisfactory) balance of risk and benefit. The general strategy of risk management using risk optimisation is to consider a number of policy or management alternatives, estimate risks and benefits associated with each, and to select the option having the most attractive outcomes. Applying this strategy is not simple and requires a number of judgments about how options are defined, and what criteria are used to rank outcomes. The key point is that risk management decisions are based on principles of risk optimisation that look to the potential impact of practices or policies on affected parties, and frame the management decision in terms of finding some acceptable balance among negative and positive impacts.[23]

Informed consent stresses the role of independent decision makers in making evaluations about the acceptability of risks. Here the task of risk management may involve much less actual assessment of outcomes, and may instead emphasise social institutions that place affected parties in a position to accept or reject risks. The most critical element in such approaches is that potential risk bearers have some means of exit, some opportunity to 'opt out' of a risky situation. The positive role of risk management under a philosophy of informed consent may be confined to ensuring that potential risk bearers have options and to providing information so that their choices can be well informed. As with risk optimisation, there can be many ways in which the basic orientation of informed consent is operationalised. Different parties face different risks. For example, consumers may be most focused on personal autonomy, favouring labels for genetically modified foods, while commercial growers may be more focused on economic gains or losses and be more concerned about access to international markets. One party's ability to opt out in such a situation may be perceived as foreclosing another's opportunity to accept risk.[24]

The precautionary principle has emerged as a prominent alternative to risk optimisation in debates over genetically modified foods. Like risk optimisation, it is an approach that places emphasis on the evaluation of outcomes. However, proponents of this approach see it as placing more emphasis on uncertainty and the reversibility of damages than do many applications of risk optimisation. When the potential hazards associated with a practice are highly uncertain, or when they are perceived as irreversible, a precautionary approach advises against attempts to quantify these risks or to weigh them against potential benefits and advises precautionary action. Opponents of using the precautionary principle as a regulatory framework have argued that to the extent it can be meaningfully applied in risk management, it is already reflected in risk management.[1]

One broad implication is that both risk optimisation and the precautionary approach tend to construe the role of public involvement as one of advising risk managers about important values, and providing information that will be useful in weighing trade-offs between risk and benefit. Risk communication then becomes interpreted as a mechanism for eliciting this information, and for managing public reactions so that mistrust and misinformation do not subvert the goals of risk management. In contrast, an informed consent approach may tend to see affected parties as key decision makers. Risk communication may then be seen both as facilitating affected parties' decision making, and as a mechanism for coordinating and reconciling differences of perspective and orientation that may exist among these groups. A result may be 'successful' in terms of informed consent because key decision makers have determined the result, but may be 'unsuccessful' from a perspective that emphasises the trade-off between beneficial and harmful impacts. In recent years, attempts to emphasise informed consent have been especially sensitive to the rights of those affected parties that operate from positions of economic or political disadvantage.

It may be difficult to apply criteria of informed consent to many situations involving ecological risk. When hazards are widespread or involve harm to ecosystem processes it is difficult to imagine how key risk bearers could be identified or their consent could be obtained in any meaningful way. However, the Mexican case discussed above illustrates a situation in which a risk management approach stressing informed consent could lead to very different processes for involving Mexican smallholders, as well as other members of the public. Although critics of biotechnology have stressed the precautionary principle, it is at least possible that decision makers utilising precaution might still develop a risk management plan that involves little consultation with affected parties. In contrast, an emphasis on informed consent would bring up the importance of involving risk bearers in the decision process more forcefully.

7 Issues in Governance with Respect to Risk

In addition to this catalogue of categories for risks that may be relevant to transgenic maize, management of risk can be complicated by a host of factors rooted in the nature of governance, and in the disparate and often unequal power, information and ability to displace risk experienced by affected parties. Such issues can often be neglected in technical discussions of risk when risk assessment is conducted purely as an exercise in decision support. In the most typical regulatory situation, a regulatory agency operates with specific legislative authority to manage a specific class of risks as stipulated in the authorising legislation. Thus, for example, the US Toxic Substances Control Act stipulates a fairly specific set of human health hazards that are to be the focus of risk management for hazardous substances. In such circumstances, important judgmental elements of risk management have been incorporated into the authorising legislation. Within such a situation, regulators have been given a specific mandate to base decisions on specific criteria and not others. Not only do such circumstances make the selection of hazards and exposures relatively straightforward, they also create a situation in which risk analysis can be perennially revisited and revised as the decision process matures or as the problem identification changes. However, flexibility must be created for the consideration and management of newly discovered risks.[5]

Increasingly, however, risk assessment is being expected to play a much less well-defined role in shaping public decision-making. For example, risk assessment and risk management are mentioned in international conventions where there is no clear agreement about the nature of relevant risks or the balance between criteria of consent and optimisation in the management philosophy.[25] In such circumstances, a risk analysis report is completed and the document is made public. This means that any gaps or omissions from the assessment can have enormous consequences. In the normal regulatory setting where risk analysis serves as decision support, regulators simply go back and collect new information. In a setting where risk assessment is put forward to structure a debate or to provide a rationale for a particular course of action, failure to note a category of risk that is extremely important to one group of affected parties can either bias the results unfairly, or can undermine the credibility and legitimacy of the entire effort to base decisions on a scientific assessment of risks. Such sources of significant (though usually unintended) bias may arise when technical experts more accustomed to analysing risk as a form of decision support are enlisted to prepare documents that have a more ambiguous and less easily controlled function.

The situation with respect to evaluating the impact of transgenic maize on Mexican farmers and the Mexican environment is an instance in which the specific role of risk assessment is unclear. While there are reasonably clear public mandates for regulation of hazards associated with public health, intervention to prevent erosion of cultural values has a much weaker basis in law and governmental practice. Arguably, ecological hazards occupy a position somewhere between the two. This article itself, as well as any subsequent empirical risk analysis that might be done to fill in gaps in our knowledge, may reflect existing practices utilised in risk analyses designed for much narrower advisory purposes far more than it reflects a complete or balanced approach to the scientific and cultural dimensions of ecological risk. Such would be a limitation of any similar review, and this fact should simply be noted.

References

1. National Research Council, *Environmental Effects of Transgenic Plants,* National Academy Press, Washington, DC, 2002.
2. National Research Council, *Risk Assessment in the Federal Government,* National Academy Press, Washington, DC, 1983.
3. F. Lynn, P. Poteat and B. Palmer, The Interplay of Science, Technology, and Values in Environmental Applications of Biotechnology, *Policy Studies Journal,* 1998, **17(1)**, 109–116.

4. P.B. Thompson, Value Judgments and Risk Comparisons: The Case of Genetically Engineered Crops, *Plant Physiology,* 2003, **132,** 10–16.
5. C. Sunstein, *Risk and Reason: Safety, Law and the Environment*, Cambridge University Press, Cambridge, 2002.
6. J. Carpenter, A. Felsot, T. Goode, M. Hammig, D. Onstad and S. Sakula, *Comparative Environmental Impacts of Biotechnology-Derived and Traditional Soybean, Corn and Cotton Crops,* Council for Agricultural Science and Technology, Ames, IA, 2002.
7. R.L. Brown, Z-Y. Chen, A. Menkir and T.E. Cleveland, Using biotechnology to enhance host resistance to aflatoxin contamination, *African Journal of Biotechnology,* 2003, **2(12),** 557–563.
8. Environmental Protection Agency, Bt *Plant-Pesticides*, Biopesticides Registration Action Document, 2000. *http://www.epa.gov/scipoly/sap/2000/october/brad5_benefits_corn.pdf*
9. G.E. Rossmiller, *Agricultural Sector Planning–A General System Simulation Approach,* Michigan State University Press, East Lansing, MI, 1978.
10. National Research Council, *Understanding Risk: Informing Decisions in a Democratic Society*, National Academy Press, Washington DC, 1997.
11. A. Snow and P. Moran Palma, Commercialization of Transgenic Plants: Potential Ecological Risks, *BioScience,* 1997, **47(2),** 86.
12. P. Kareiva and M. Marvier, An Overview of Risk Assessment Procedures Applied to Genetically Engineered Crops, *Incorporating Science, Economics, and Sociology in Developing Sanitary and Phytosanitary Standards in International Trade: Proceedings of a Conference*, Board on Agriculture and Natural Resources, National Research Council, National Academy Press, Washington DC, 2000, 231–238.
13. L.L. Wolfenbarger and P.R. Phifer, The Ecological Risks and Benefits of Genetically Engineered Plants, *Science,* 2000, **290**, 2088–2093.
14. M. Mellon and J. Rissler, *Gone to Seed: Transgenic Contaminants in the Traditional Seed Supply,* Union of Concerned Scientists, Cambridge, MA, 2004, 36.
15. M. Nestle, *Safe Food: Bacteria, Biotechnology and Bioterrorism,* University of California Press, Berkeley, CA, 2003.
16. D. Quist and I. H. Chapela, Transgenic DNA introgressed into traditional maize landraces in Oaxaca, Mexico, *Nature,* 2001, **414**, 541–543.
17. M. Bellon and J. Berthaud, Transgenic Maize and the Evolution of Landrace Diversity in Mexico: The Importance of Farmers' Behavior, *Plant Pathology,* 2004, **134**, 883–888.
18. P.B. Thompson, *Food Biotechnology in Ethical Perspective,* Chapman and Hall, London, 1997.
19. M. Bellon and S. Brush, Keepers of the Maize in Chiapas, Mexico, *Economic Botany,* 1994, **48**, 196–209.
20. M. Douglas and A. Wildavsky, *Risk and Culture,* University of California Press, Berkeley, CA, 1982.
21. D. Barling, H. de Vriend, J.A. Cornelese, B. Ekstrand, E.F.F. Hecker, J. Howlett, J.H. Jensen, T. Lang. S. Mayer, K.B. Staer and R. Top, The Social Aspects of Biotechnology: A European View, *Environmental Toxicology and Pharmacology,* 1999, **7**, 85–93.

22. M.W. Bauer and G. Gaskell (eds), *Biotechnology: the Making of a Global Controversy,* Cambridge University Press, Cambridge, UK, 2002.
23. T. Deitz, R.S. Frey and E. A. Rosa, Risk, Technology and Society in R.E. Dunlap and W. Michelson (eds), *Handbook of Environmental Sociology*, Westview Press, Boulder, CO, 2002, 329–369.
24. P.B. Thompson, Why Food Biotechnology Needs an Opt Out, in *Engineering the Farm: Ethical and Social Aspects of Agricultural Biotechnology,* B. Bailey and M. Lappé (eds), Island Press, Washington, DC, 2002, 27–44.
25. National Research Council, *Incorporating Science, Economics, and Sociology in Developing Sanitary and Phytosanitary Standards in International Trade: Proceedings of a Conference*, National Academy Press, Washington DC, 2000.

GM Pest-resistant Crops: Assessing Environmental Impacts on Non-target Organisms

A.N.E. BIRCH AND R.E. WHEATLEY

1 Introduction

Human impacts on the biosphere now control or strongly influence many major facets of global eco-system functioning, including agriculture.[1] Human ecological impacts have enormous evolutionary consequences, by exerting strong natural selection from introduced technologies. This frequently accelerates the rate of evolutionary change in species which are important to humans as agricultural pests, disease organisms and providers of important 'ecological services' (*e.g.* species acting as biocontrol agents, crop pollinators, soil nutrient cyclers, *etc.*). Introduced technologies can result in cycles of perceived success and under-achievement. For example, Paul Muller's discovery that DDT killed insects in 1939 won him the Nobel prize in 1948. However, by 1948 resistance to DDT had already been reported in house flies. By the 1960s, mosquito species resistant to DDT prevented worldwide eradication of malaria. By the 1990s more than 500 pest species have evolved resistance to at least one insecticide, often within ten years of commercial introduction.[2]

The science of entomology has the unique task of understanding the biology of some 80% of known species on Earth, of which 85% remain uncollected and undescribed. Pests of agricultural crops destroy or eat food worth $5 billion per year (averaging 20–30% losses of pre- and post-harvest crops, enough to feed 1 billion people per year), despite annual pesticide applications valued at $6 billion per year. As vectors of animal and human diseases, insects cause huge economic impacts; insect-borne diseases such as malaria weaken or kill 200 million people per year. In sharp contrast to these negative impacts, insects and other invertebrates from the phylum *Arthropoda* (insects, spiders, crustaceans, millipedes, centipedes and others) also provide invaluable 'ecological services' to agriculture.[3] About one-third of the world's crop plants depend on insects for pollination, estimated to be worth about

Issues in Environmental Science and Technology, No. 21
Sustainability in Agriculture

$117 billion per year. The overall value of natural biological control (termed 'bio-control') is about $400 billion per year, mostly provided by insects. The value of soil nutrient cycling is over $3 trillion per year. Insects and related arthropods can compose half the animal biomass in some tropical forests.[4]

The introduction of high-yielding crop varieties in combination with high inputs of synthetic fertilisers, pesticides and the use of irrigation systems, transformed agriculture in Asia and Latin America, later spreading to Africa. However, in several instances the initial rapid successes of the 'green revolution' decreased quickly. A major factor in this under-achievement was the over-use of non-selective pesticides on pest-susceptible crop varieties, which increased the rate of evolution of insecticide-resistant pests, removed important biocontrol agents from agro-ecosystems and led to resurgences of primary and secondary pests. This resulted in further increases in pesticide use (the so-called 'pesticide treadmill' effect). Such effects are estimated to cost $3–7 billion in the USA alone, due to increased crop losses and increased pesticide use.[5] This pesticide problem began in the 1940s and peaked in the 1970s, but still continues to influence agricultural practices today. Similar effects have been observed around the world after the introduction of pest-resistant crop varieties through conventional breeding.[6] After the initial successes exemplified by resistance to Hessian fly in wheat in the early 1960s, plant breeders constantly struggled to stay ahead of insect pest evolution (populations of pests able to overcome genetically based pest resistance or applied synthetic pesticides), after exerting strong selection pressures on pest populations. Selected biotypes (adapted populations) of agricultural pests can often overcome single gene-based pest resistance faster than plant breeders can introduce and release a variety with a different resistance gene.[7,8]

Since the 1980s we have entered the 'gene revolution', using genetic modification (GM) to engineer novel genes into crops, tailoring them to address our current crop production and crop protection problems. The control of pests is still a major and increasingly important issue, since many existing pesticides (particularly broader spectrum products like organophosphates) are being withdrawn from global markets on environmental grounds. Genetic resistance in crops to pests and diseases is often overcome before new genes can be bred into commercially available varieties. The commercial introduction of insecticidal bacterial toxins (now renamed 'plant-incorporated protectants' in USA) from *Bacillus thuringiensis* (*Bt*) in GM crops since 1996 has provided a moderate to high level of resistance against a range of damaging 'target' pests in some major crops, particularly for cotton and maize. During the eight-year period from 1996–2003, the global area of GM crops has increased 40-fold, from 1.7 million hectares in 1996 to 81 million hectares in 2004.[9] Currently about 29% of total global areas for soybean, cotton, maize and canola are GM, up from 25% in 2003. Of this global area of GM crop cultivation, 15.6 million hectares (19%) was planted with insecticidal *Bt* toxin-expressing crops in 2004. Combined *Bt* and herbicide tolerant (HT) GM crop varieties have also increased substantially, illustrating the increasing trend for 'gene stacking' in current generations of GM crops. Gene stacking may also involve combining different insecticidal toxin types (*e.g.* combining more than one *Bt* toxin type in *Bt* cotton; *Bt* combined with other types of insecticidal or anti-metabolic proteins including VIPs, lectins, protease inhibitors, insect toxins and RNA interference (RNAi) of host-pest recognition genes, *etc.*, in future GM crops).

Lessons from the 'Green Revolution' for the 'Gene Revolution'

What lessons can we learn from the 'green revolution' as we accelerate further into the 'gene revolution' using a wider range of GM crops as well as other recent advances in biotechnology, including molecular marker-assisted breeding and genetic manipulation of plant metabolic pathways? Many questions arise amongst scientists, consumers, environmentalists, GM crop regulators and governmental policy-makers as we look from the recent past of the green revolution to the current situation and near future. GM crop technology is already being widely cultivated (*i.e.* more than 50,0000 hectares per country) in the USA, Canada, Brazil, Argentina, China, South Africa, Australia, India, Uruguay and Romania. A number of important ecological, socio-economic and world trade questions are emerging as we reflect on the short history of expanding GM crop usage to date: How will developing countries like Mexico, Honduras, Colombia, Philippines, Indonesia and Bulgaria respond to more recent introductions of GM crops? Will Europe follow the lead of Spain and Germany (currently < 0.05 million hectares of only one GM crop, *Bt* maize) and adopt GM technologies more widely in the future? Can we select optimal combinations of GM and non-GM technologies and develop durable deployment strategies for more sustainable forms of agriculture and crop protection? Can we stay ahead of co-evolving pests and diseases in the dynamic evolutionary arms race between plants and a diverse array of evolving and adapting pests and diseases? Will future deployment of pest-resistant crops (GM or conventional) within integrated pest management (IPM) help us to slow down this evolutionary arms race?
Wu and Butz[10] review lessons from the green revolution relevant to the use of GM crops in the gene revolution. They identify key features of agricultural 'revolutions' from a historical perspective. These include:

- Giving incentives to farmers for increased productivity (farmer benefits).
- Improving crop and food production efficiency (environmental and consumer benefits).
- Willingness to adapt culturally and economically to new technologies (by providing clear benefits to farmers, consumers and governments).
- Cooperation between providers, regulators and users of the technology.
- Sustainable technology-driven movements, eventually without public or private subsidies.

The authors also state many similarities between the green and gene revolutions, but identify some key differences, which we expand on:

- The science and technology required to create GM seeds is far more complex than previously using conventional plant breeding.
- GM seeds are created largely through private enterprise, although this is slowly changing and public sector GM crops are starting to reach the testing phase in developing countries like China, Kenya and the Philippines.
- The political climate in which agricultural science influences the world has changed dramatically in the last decade, aided by global and instantaneous

e-media communication. The latter has polarised the GM debate far beyond the reasonable zone most scientists expect to operate in, where more or less quantified 'shades of grey' in scientific knowledge are the norm.

- Food-related technology and the general status of science in society has been damaged by several UK and European food chain-related scares in the last decade including bovine spongiform encephalopathy (BSE or 'mad cow disease'), bacterial pathogens including *Salmonella* and *E. coli* 0-*157* and the viral 'foot and mouth disease' in livestock. Although most are animal-related problems, they have affected public perception of food safety, agriculture, confidence in science and environmental impacts of new technologies in general.
- European societies have become more risk-averse as a result, and have put more pressure on scientists to adopt the 'precautionary approach' (*i.e.* state regulators proving new technologies are acceptably 'safe' before release). Many countries (111 to date) have now signed and ratified the Cartagena Protocol on Biosafety. This international agreement incorporates the precautionary approach as a scientific foundation. However, several major GM crop-producing countries like USA or Australia have either not signed or not ratified this agreement.

All of these points influence ways in which farmers, consumers, the media, GM regulators and governmental policy-makers view the relative risks and benefits of GM crops, compared with existing methods for safe crop production and crop protection.

2 Possible Risks and Benefits of Pest-resistant GM Crops for Above and Below Ground Agro-ecosystems

Above-ground Interactions with Agro-ecosystems

Generally, risks and benefits of pest-resistant GM crops have been well aired over the last decade, particularly since the commercial introduction of *Bt*-expressing crops in 1996. To date mainly *Bt*-expressing crops are used in commercial practice, although lectin and protease inhibitor-expressing crops are being tested in developing countries like China, sometimes in combination with *Bt* toxins as stacked gene events. Despite nearly ten years of increasing commercial use, many scientific uncertainties and controversies remain among scientific experts and GM crop regulators concerning *Bt*-expressing and other pest-resistant GM crops. Some of the controversy resides in the variable usage of risk assessment terminology, which is not well standardised between countries. The UK Department of the Environment, Transport and the Regions define 'hazard' as the intrinsic properties of an activity, substance or organism (*e.g.* a particular GM crop and its commercial use in agriculture) that may (potentially) cause harm. Risk is a more complex concept and is thus often confused with hazard in the GM debate. According to Defra in the UK, 'risk' combines the likelihood of the 'hazard' being realised (*e.g.* degree of exposure to a *Bt* toxin by a 'non-target' organism) together with the magnitude of the likely consequences of exposure (*i.e.* how serious will the effect be in ecological and socio-economic terms?).

In the UK, as a ratified signatory to the Cartagena Protocol in 2004, this has led to a general approach for risk assessment of GM crops which currently includes:

- Identifying intrinsic properties that may cause harm to humans, livestock, wildlife and/or the environment.
- Estimating the likelihood of those effects occurring, under the proposed conditions of use of the GM crop and its crop management system.
- Estimating the likelihood and magnitude of the harm that may arise if the effects (potential harm) occur.
- On this basis, evaluating the overall risk of the GM crop in the receiving environment (*e.g.* a particular agricultural region or country).

Currently in the UK this risk assessment approach does not include comparisons with current alternatives (*e.g.* best current practice for controlling particular pest/crop combinations in conventional or organic farming systems), nor does it include a cost–benefit analysis of the new technology versus current alternatives, as advocated by many scientists.[11] Both these 'add-ons' to GM crop risk assessment in the UK are considered by many stakeholder groups in society to be desirable but would require more detailed and accurate data. This should be collected over several years and locations before becoming useful to stakeholders in the GM debate, and is therefore moderately expensive in terms of time and research funding needed.

Potential Risks of Cultivating Pest-resistant GM Crops

These include:

- Pest resistance (*via* co-evolution in pest populations) to the GM insecticidal toxin could nullify the efficacy of expressed *Bt* toxins, *via* increased selection pressure on target pest populations (*e.g.* by growing large cultivation areas of *Bt* crops expressing the same *Bt* toxin type within a geographical area occupied by pests able to feed on more than one *Bt* crop type).
- Adverse non-target effects on beneficial non-target species, above- or below-ground, on diverse taxa that provide 'ecological services' or that are of special conservation value (*e.g.* regionally important butterfly species). Such adverse effects can be 'direct' (*e.g.* adverse effects caused directly by the *Bt* toxin, impacting on the ecological fitness of a non-target organism following exposure) or 'indirect', *via* trophic interactions (food webs) involving the GM crop in its agro-ecosystem (*e.g.* by reducing prey quality and or quantity of herbivores feeding on a *Bt* crop to a predator at the next trophic level). Both direct and indirect effects, or the combined effects of both, can be ecologically important as lethal and sub-lethal selection pressures, potentially accumulating over multiple insect generations and several cropping seasons.[12]
- Adverse effects on agriculturally important 'ecological functions' rather than specific non-target species, some acting over the many seasons or years (*e.g.* soil decomposition and nutrient cycling rates effects on soil agro-ecosystems if continuous monocultures of *Bt* crops are grown).

- Lack of sufficient crop protection efficacy *versus* pest complexes (*i.e.* several pest species attacking a crop, differing from region to region and season to season). These pest complexes can attack the crop season-long, illustrating that *Bt* crops are not 'magic bullets' for use on their own, as they only control certain pest species from the total local pest complex.
- 'Knock-on' effects of GM crop management systems when replacing existing systems (*e.g.* the introduction of a GM crop could cause primary or secondary pest resurgences after altered pesticide use, reduction of crop genetic diversity, loss of local/traditional cropping methods previously using IPM).
- 'Unintended' ecological and nutritional effects on trophic interactions (*e.g.* herbivores, beneficial predators, parasitoids and pollinators for ecological food webs) as a result of the transformation (genetic engineering) process (*e.g.* altered primary or secondary plant metabolites important as semiochemicals (signals used by insects and other herbivores) and defence compounds selected in plants over evolutionary, ecological and plant breeding timescales).
- Unexpected genotype × environment (g×e) interactions, causing reduced efficacy of the GM crop (*e.g.* decreased pest resistance, reduced yield) or unintended, adverse effects on beneficial insects *via* trophic interactions (*e.g.* involving the GM crop, its herbivorous pests, predators and parasitoids of locally abundant pests). These complex interactions are often influenced by biotic or abiotic stress factors not always encountered during the development and testing of the GM crop (*e.g.* high temperatures combined with low water availability can reduce *Bt* toxin expression and efficacy against target pests, because stress factors apparently alter plant metabolism and then the level of *Bt* toxin production).
- Yield drag effects, supposedly caused by the metabolic burden of producing novel insecticidal protein(s), but possibly also due to unintended changes in plant metabolism and physiology arising during the genetic transformation process.[13,14]
- Pest-resistance genes could spread, *via* gene flow, to other non-GM crops (environmental and seed contamination issues) or to local landraces and sexually compatible wild relatives of the GM crop (regional biodiversity issues, *e.g.* teosinte maize landraces in Mexico crossing with *Bt* maize). Introgression of novel pest-resistance genes ('genetic pollution') could alter the ecological fitness of the local plant landrace or wild species and then change the dynamics and balance of biodiversity in environmentally sensitive regions, such as centres of genetic diversity for crop plants or in nature reserves. If non-GM plants acquire insect resistance from GM or conventional crops they could damage food webs which are dependent on insects feeding on previously 'non-toxic' wild plants or crops.

Potential Benefits of Cultivating Pest-resistant GM Crops

These include:

- Greater efficacy against some key 'target' pests (*i.e.* those controlled by the GM toxin), many of which are difficult to control using synthetic insecticides (*e.g.* stem borers and rootworms which feed inside maize or in the soil) or which

require multiple insecticide applications because of high pest pressure from several different pest species throughout the growing season, *e.g.* the pest complex of cotton in Australia, which typically includes several lepidopteran species, several aphid species with associated viral diseases, a true bug, several whitefly species, jassids (plant hoppers), weevils, several thrips species, mirids and spider mites.
- More stable yields and economic benefits when the feeding pressure from the 'target pest(s)' is high and when growing conditions favour the GM crop variety (the GM crop 'event').
- By design, greater specificity of *Bt* toxins than most broad-spectrum insecticides applied by sprays or granules. *Bt* toxins are generally considered to be specific to insect orders, *e.g.* activity against *Lepidoptera* or *Coleoptera*, with several insecticidal *Bt* toxins being more effective against specific pest genera or species within orders. Some insecticidal proteins used in GM crops, including protease inhibitors and lectins[15,16] are much less specific than currently used *Bt* toxins and so are more likely to show environmental problems on non-target species previously caused by broad spectrum insecticides.
- Potentially less need to spray insecticides, at least against the target pest(s) which are controlled by the *Bt* crop. This potentially offers: a) greater level of protection to the crop plant in certain situations (i.e. when the target pest pressure is high but pressure from non-target pests is low), b) reduced pollutant inputs to the environment (*i.e.* reduced spray drift, pesticide residues in soil and groundwater, c) reduced use of diesel fuels during application, d) reduced spray exposure to the operator and certain benefits to the consumer (less pesticide residues), and e) less insecticide impact on non-target organisms, *via* any resulting reductions in insecticides that were formerly used in current practice growing a pest-susceptible variety.
- However, other, 'non-target', or secondary pests which are not controlled by the GM toxin (*e.g.* sucking insects like aphids, whiteflies and plant hoppers not controlled by toxins expressed in *Bt* crops) still need to be controlled by applications of synthetic pesticides. This can result in pesticide reductions, no change in levels of use, or increased pesticide use on some *Bt* crops, depending on local pest pressures, local management practices, economic factors and environmental conditions.

Evidence from Small-scale Experiments Using 'Contained' Laboratory or Glasshouse Tests

Interactions between pest-resistant crops (GM or conventional) are generally complex, involving food webs across multiple trophic levels.[3,11,12,17,18] Effects of a pest-resistant crop detected at a small scale of experimentation can be:

- 'Direct' (*i.e.* due directly to the expressed toxin or resistance trait on the consumer).
- 'Indirect' (*i.e.* due to one or more secondary impacts on fitness of a herbivore or on ecologically linked organisms at higher trophic levels in food webs, *e.g.* reduction in host availability or quality for predator or parasitoid).

- 'Knock-on' (*i.e.* due to longer-term impacts of the cultivation of the GM crop and implementation of the GM crop's management system). These effects may be detected in extended small-scale studies but are more likely to be detected in multi-season field experiments, using regionally appropriate cultivation and crop management systems.

Groot and Dicke[17] review impacts of *Bt* toxins on a range of parasitic and predatory natural enemies, following direct exposure (*via* purified toxins or plant products containing *Bt* toxin) and indirect exposure (*via* a herbivorous host insect which had consumed a *Bt* toxin from an artificial diet, from a plant sprayed/dipped with *Bt* toxin or from a *Bt*-expressing crop). A wide range of effects were reported on a non-uniform and diverse set of ecological fitness parameters (*e.g.* development rate, fecundity, mortality) and behavioural/physiological parameters (*e.g.* attraction, parasitism rate, predation rate). Of the 58 studies reviewed by Groot and Dicke (24% from field-based studies, 76% from laboratory-based studies), 19 of the reviewed studies (33%) showed negative impacts on natural enemies, 6 (10%) showed positive impacts and 33 (57%) showed no or neutral/non-detectable effects. These results broadly reflect results studying impacts of conventionally bred pest-resistant crops, where 38 (67%) of 56 studies showed positive effects, 8 (14%) showed negative effects and 10 (18%) showed no statistically significant/neutral effects on natural enemies. The authors conclude that natural enemies are good indicators of potential ecological impacts of GM crops as they are economically important and belong to the third trophic level in food webs. In a more recent review, covering 44 laboratory-based studies on the potential impacts (positive, neutral or negative) of pest-resistant GM plants on natural enemies, Lovei and Arpaia[19] found that 30% of these studies showed negative effects on predators and 40% of studies showed negative effects on parasitoids. They also point out that only 18 species of predators and 14 species of parasitoids have been tested to date, most of which were tested only in a few experiments. Certain natural enemy groups (*e.g.* braconid wasps) or single species (*e.g.* the green lacewing, *Chrysoperla carnea*) have attracted much research effort, while representatives of other whole natural enemy orders which are important in biocontrol (*e.g.* Diptera) have been largely overlooked to date. They and other authors[12,18,19] criticise laboratory-based studies as not necessarily representing 'worst case scenarios' and for being ecologically unrealistic. This calls into question the value of small-scale experiments in predicting large-scale, longer-term effects, particularly involving sub-lethal and behavioural effects on fitness parameters and population dynamics over multiple generations and growing seasons.

Potential benefits of laboratory- or glasshouse-based, small-scale experiments for evaluating potential impacts of GM crops on non-target organisms include:

- They are generally easier, faster and cheaper to perform than field-based tests.
- Environmental factors (temperature, light, humidity) are often kept constant, so reducing experimental variability and environmental interactions.
- Indicator species often derived from eco-toxicological studies on pesticides can be reared under laboratory conditions.

- Food webs operating in natural and agro-ecosystems can be greatly simplified, to allow greater precision in understanding one part of a more complex set of dynamic interactions.
- Demonstration of no effect in the laboratory or glasshouse could negate the need to carry out additional, more expensive testing involving semi-field and field experiments, further up the line in 'tiered tests' for risk assessment of GM crops, as proposed by some scientists.[20]

However, many scientists consider that the potential disadvantages of small-scale laboratory or glasshouse tests for predicting harm of GM crops to non-target organisms need to be carefully considered and taken into account. Such disadvantages of small-scale and short-term biosafety tests include:

- The simplified experimental design, using pre-selected 'indicator species' does not accurately reflect ecological reality under field conditions. In agricultural reality, biotic and abiotic factors affecting plants are variable. In addition, GM crop genotype × environment interactions are likely, are often unpredictable and can operate at several trophic levels, *e.g.* affecting first trophic level (*i.e.* GM crop together with other associated vegetation in the agro-ecosystem), second trophic level herbivores ingesting GM crops (*i.e.* target pests, non-target pests, non-pest species), third trophic level natural enemies of herbivores ingesting GM crops (*i.e.* predators and parasitoids), fourth trophic level super-parasites or predators of natural enemies and below-ground interactions affecting the soil ecosystem (*i.e.* multiple soil dwelling taxa and multiple trophic levels). Each trophic level is likely to be affected over a wide range of spatio-temporal scales, from the intra-cellular, inter-cellular, organism levels, to field ecosystem and landscape levels. The spatial scale effect also has the potential for different degrees of delayed impacts, across timescales ranging from hours (if acute toxicity or obvious effects on behaviour and physiology) to days, weeks or seasons.
- For most ecologically realistic exposures involving pest-resistant GM crops, the insecticidal GM toxin concentration will be low to moderate but exposure time will be long (season-long). Therefore, testing for sub-lethal effects following chronic exposure of ecologically realistic and potentially variable toxin doses over several weeks would seem much more realistic and useful in environmental risk assessment than conducting very short term, 'acute' toxicity tests using unrealistically high doses of *Bt* toxin. Such acute toxicity tests are often taken directly from eco-toxicological methods used for pesticides and cited as simulated 'worst case scenarios'. However, if future generations of insecticidal crops use 'over-expression' (*e.g.* chloroplast transformation) or other technologies to increase the level of *Bt* toxin expression in the GM crop by an order of magnitude or more, then such acute toxicity testing on non-target organisms could become more ecologically informative.[21]
- Small-scale experiments, attempting to mimic ecological reality to different extents, can only indicate potential hazards (not risks), under artificial conditions. Risk (involving estimates of the likelihood/probability and in some definitions

also the likely extent of harm) can only be assessed under a realistic range of agricultural conditions representing regional or country-specific growing conditions.

- Purified GM *Bt* toxins when tested at very high exposure rates for short durations (typically a few days) do not reflect environmental reality for toxin dose, route of exposure nor duration of exposure. The lessons learnt from DDT and the green revolution appear to have been forgotten by some scientists using eco-toxicological tests designed for pesticides to test effects of a growing plant which interacts dynamically with its environment and ecosystem.
- The purified GM toxin tested (called a 'surrogate' GM product if purified from non-plant sources) may be structurally and functionally different from that expressed in the intact GM plant when growing in its normal environment, so may differ in bioactivity when compared to the surrogate toxin tested). Generally 'surrogate GM proteins' derived from bacteria are used in place of the plant-produced protein because the former are easier to produce. However, it is likely that bacteria and plants will not make identical protein when transformed with the same gene, due to differential processing involving glycosylation and other secondary modifications, or alterations caused by point mutations.[21]
- Many bio-active plant metabolites show non-linear or 'hormetic' dose-response curves as demonstrated by Calabrese and Baldwin.[22] For example, plant metabolites can be pro-biotic or stimulant to test species at low concentration but switch to becoming toxic or deterrent to the same species at higher concentration ranges. Conversely, some bio-active molecules are more effective at low concentrations than at higher concentrations (*e.g.* some elicitors of plant defence responses). Hormetic dose responses are now well documented in risk assessment and challenge our long-standing assumption that low-dose responses to toxins can be extrapolated from high to low doses with adequate precision and accuracy.
- Testing purified GM toxins in simplified artificial diets for insects (diets which are generally sub-optimal compared with the actively growing host plant) or testing excised plant GM material does not reflect ecological reality, where a complex array of primary plant metabolites and constitutive and inducible secondary plant metabolites are produced. These plant compounds interact in complex ways to determine the suitability of the plant as a host for feeding, development and reproduction. Many primary and secondary compounds act additively or synergistically in plant defence mechanisms and are likely to interact with an expressed GM toxin. These well defined interactions involving plants (from hundreds of chemical ecology studies) cannot be measured accurately using artificial diets or excised plant tissues.
- Testing synthetic pesticides for biosafety (the typical eco-toxicological model used) generally involves not only the 'active ingredient' but also the 'formulation' in which it will be applied (sticking and wetting agents, synergists, *etc.*), over a wide range of doses and environmental conditions. We consider the actively growing GM plant which is expressing an insecticidal toxin to be equivalent to the 'formulation' tested routinely in eco-toxicological evaluation of synthetic pesticides. Testing just the purified GM toxin (the 'active ingredient') at high concentration in an artificial diet, or testing excised GM plant material

which is metabolically compromised does not mimic conventional pesticide testing developed over many years nor environmental reality.
- Biosafety tests using artificial diets or excised plant material preclude or minimise the sensitivity of studies to detect any unintended effects in the GM plant arising during the transformation process (*e.g. via* pleiotrophic effects, epistasis or insertional mutagenesis). Such effects are often caused by the random nature of gene insertion(s) or parts of genes into plant DNA.[21,23]
- Unintended effects detected in GM and conventional crops can accidentally modify plant metabolic pathways involved in production of 'infochemicals' and plant defence compounds.[13,17,24] Many insects use these plant chemicals to detect suitable host plants (*e.g.* second trophic level herbivores and pollinators) and prey (*e.g.* third trophic level predators and parasitoids).

Field-based Evidence of Pest-resistant GM Crop Impacts (Pre- and Post-release Studies)

Because of the cost and complexity of field-based studies of GM crops, inevitably compromises have to be made.[25] These compromises often include:

- Size and duration of the field experiment.
- Number of species which can be studied from the local food web (if previously characterised).
- Lack of baseline data to determine any future change due to introduction of the GM crop in a region.
- Replication of sampling effort, affecting the statistical power of detecting statistically significant effects.
- Number of time points for measurements during the growing season, or over multiple seasons (longer-term effects modified by rotations and other farmer practices).
- Number of GM and conventional varieties included.
- Comparison of different treatments (*e.g.* +/– pesticides) and crop management systems (conventional *versus* IPM *versus* organic) in different studies, making generalisations very difficult.
- Full assessment of the environmental, economic and social impacts of the new technology.

In a review of 53 field studies[25] mainly in the Northern Hemisphere, of GM crop impacts on arthropod natural enemies (*e.g.* predators and parasitoids of agricultural pests) between 1992–2004, most studies were only conducted on a small scale (< 1 ha plots) and many were considered by the authors to have insufficient replication over space and time to make strong conclusions on non-target impacts. Impacts are measured using a wide range of parameters and on single species to species complexes or various indices of biodiversity. Thus, from existing publications, generalisations and broader predictions on field, regional or between-country scale impacts of growing GM crops like *Bt* maize and cotton are difficult and subject to scientific uncertainty and ongoing debate.

Despite these limitations, several studies and reviews of pest-resistant GM crops (*Bt* crops) have now been published. We will not include discussion of the results from the UK 'Farm Scale Evaluations' (FSE) of GM oilseed rape (winter and spring types, maize and sugar beet) because the GM trait and associated management systems concerned herbicide tolerance (HT), not pest resistance. In Groot and Dicke's review[17] covering 14 field-based studies of *Bt* crops or *Bt* applied as a spray to conventional crops, only 2 field studies (14%) showed negative effects on non-target natural enemies, 9 field studies (62%) showed no detectable or neutral effects on natural enemies and 3 field studies (21%) showed positive effects on natural enemies (*i.e.* benefits *versus* the comparator used in each field study). The effects of sample size, statistical power, and duration of study (multi-season effects) on the interpretation of results were not analysed in this review, but are likely to be even more important in field-based studies than for simplified, small-scale studies which are conducted in the relatively controlled conditions of the laboratory or glasshouse. It has been recommended[20,26] that statistical tests are accompanied by a power analysis, to avoid committing type I or II statistical errors and to indicate the precision of detection for any measured parameter for assessing impacts on non-target species (*e.g.* natural enemies) or ecological functions (*e.g.* soil nutrient cycling).

The long-term sustainability of GM pest-resistant crops (also referred to as 'transgenic insecticidal cultivars' or TICs) has been a topic of scientific debate ever since their commercial introduction.[6,27] These concerns underpinned the development of the 'high dose' (high *Bt* toxin expression)/'refuge' (regulated areas of pest-susceptible cultivars planted near *Bt* crops to reduce selection pressure for resistance to introduced *Bt* toxins in pest populations). In a retrospective web-based article entitled 'Insect resistance to *Bt* crops: lessons from the first seven years'[28] several interesting points are made. Surprisingly, after seven years of large-scale planting of *Bt* crops, pest resistance to *Bt* crops in the field has not yet been reported,[29] although several cases were reported under laboratory conditions using purified *Bt* toxin or in the field against *Bt* spray mixtures. The cultivation of *Bt* crops represents one of the largest selection pressures for resistance in insects the world has ever seen. The authors conclude that the 'high dose/refuge' strategy has probably been important in delaying resistance, even though conditions for success of the refuge strategy are not ideal in some cases. However, they urge caution, stating that this does not preclude resistance in the future and that vigilant efforts to delay and monitor field resistance are essential.

Opinions on the environmental, social and economic impacts of *Bt* crops are wide-ranging across the world. In developed countries like the USA and Australia, GM crop-, event-, regional- and season-specific benefits have been published. For example, citing *Bt* cotton grown in the USA in 1997–1998, 'elasticity of yields' (relative change as a result of GM crop adoption) was positively increased by 21% and net returns increased by 22%. However for *Bt* corn grown in the USA over the same period net returns were decreased by 34%.[30] This analysis did show an overall reduction of pesticide use for *Bt* cotton, equivalent to 6.2% of total treatments (*Bt* maize data was not available for analysis). Most of the overall decline in pesticide acre treatments on GM crops in the USA was attributed to less herbicide use on HT soybeans. However, this is complicated because total herbicide pounds (weight of active ingredient) used in herbicides actually increased as glyphosate replaced conventional herbicides for HT GM soybean.

The changing mix of pesticides that usually accompanies adoption of specific GM crops as part of their management system complicates the analysis of environmental impacts, because toxicity and persistence of formerly used pesticides and replacement pesticides now used with GM crops varies greatly. For example, for certain HT GM crops the herbicide glyphosate replaces other synthetic herbicides that are at least three times as toxic and persist in the environment nearly twice as long, so this substitution can be viewed as having a positive environmental impact. However, over-use of glyphosate could lead to glyphosate-tolerant weeds in the longer term and thus result in a switch back to more environmentally damaging herbicides. Hence assessing impacts of the long-term management of the GM crop (the so called 'knock-on effects') is now considered by many scientists to be just as important as assessing the potential short-term hazards and risks of field cultivation.

For *Bt* crops, the economics and environmental benefits *versus* costs depend on several variable factors including regional and seasonal target pest pressures, non-target pest pressures, costs of GM seeds and pesticides and other environmental stress factors also affecting crop yields. In the USA, adoption of *Bt* maize has had a negative impact overall,[30] probably because it was 'over-adopted' in some regions and seasons where the value of protection against the main target pest, the European corn borer (ECB), was less than the *Bt* seed premium. The authors cite annual variations in ECB infestations, poor forecasts of infestation levels, increased *Bt* maize prices and 'insurance purchase' of *Bt* corn seeds by risk-averse farmers as contributory factors in the overall 'negative economic impact' of *Bt* corn for US farmers. In many ways the use of *Bt* corn in the USA to date parallels the prophylactic use of insecticides when they are used as a means of reducing risks of crop losses by paying a higher premium, even in seasons when control of the target pest(s) is not necessary.

In Australia, *Bt* cotton (tradename Ingard) expressing *Cry1Ac* endotoxin have been commercialised since 1996 and have gradually increased in use in a carefully controlled IPM scheme, up to a capped maximum of 30% of the total cotton growing area.[31] In this period Ingard cotton has effectively controlled heiolothine lepidopteran pests (particularly *Helicoverpa armigera*, the main target pest) in most seasons and locations and consequently reduced pesticide needs for controlling *Helicoverpa armigera* and *H. punctigera* by an average of 56%. This has provided some important environmental benefits in Australia. With *Bt* cotton being capped at 30% (although this cap will diminish in future with second generation *Bt* Bollgard II cotton which expresses an additional *Bt* toxin, *Cry2Ab*), it has been used selectively near environmentally sensitive watercourses and townships to maximise the benefits of pesticide reduction. Some problems have been encountered using *Bt* cotton in Australia which have re-enforced the view that *Bt* crop technology is not a 'magic bullet' to directly replace conventional insecticides. For example, the main lepidopteran pest species in Australia were found to be less sensitive to *Cry1Ab Bt* toxins than *Heliothis virescens*, the main target for Bolgard *Bt* cotton in USA.[31] Additionally, variability in performance against the target pests was found to be due to the inability of Ingard cotton to produce sufficiently high doses of *Bt* toxin later in the season, when target pest larvae were able to survive and damage the plant. The risk of survival of heterozygote target pest larvae able to overcome *Cry1Ab Bt* toxin appears to have been offset in a carefully designed insect resistance management programme, by having a huge refuge area (at least 70% of the total cotton area) and using threshold-driven application of insecticides, thus diluting selection pressure for rapid pest counter-adaptation. *Bt* resistance levels are closely monitored in field populations to prevent breakdown of the 'moderate dose/high refuge' strategy adopted for Ingard cotton. One 'knock-on' effect of using *Bt* cotton in Australia with concurrent reduction in broad-spectrum insecticides (*e.g.* endosulphan, carbamates and organophosphates) over five years, has been an increased abundance of secondary pests, particularly aphids early in the growing season. This indicates that although natural enemies of aphids and other minor pests of cotton have benefited from reduced pesticide application on *Bt* cotton, this has not been sufficient so far to offset the decrease the aphid-controlling effects of early season broad-spectrum pesticides formerly used to control a broad range of cotton pests.

Overall, the economic benefits (but not including the value of the environmental benefits) of using *Bt* cotton in Australia have been fairly neutral and variable (ranging from +\$1000 per ha to –\$1000 per ha). In the first three years of Ingard *Bt* cotton use in Australia, average costs per insecticide spray were higher than for conventional cotton, but showed little or no difference after the initial three year period. The threshold system adopted initially (two positive consecutive checks before spraying) allowed surviving pest larvae to grow larger before action was taken, then requiring more expensive, 'hard' pesticides (having more negative impacts on the environment than 'soft' pesticides). With more experience growers and consultants have adjusted their expectations of Ingard *Bt* cotton performance from mid-season onwards. It is suggested[32] that introduction of Bolgard II (expressing two different *Bt* toxins, season-long) together with manipulation of natural enemies using nursery crops and food sprays will offset the problems encountered with first generation *Bt* cotton/IPM. From two full seasons of using Bollgard II cotton in Australia it seems that this second generation GM crop has much better efficacy than Ingard *Bt* cotton and has enabled reductions of about 90% in pesticide active ingredient per ha. Most Bollgard cotton now receives only 1–2 sprays per season and about 30% of the crop receives no pesticide sprays at all. In the 2004–2005 season it is planned that 100% of the GM cotton crop will be Bollgard II (Ingard cotton will be withdrawn), representing about 60% of the total Australian cotton crop. The 30% cap on area, relative to refugia area, has been lifted because the two-gene Bollgard II product is considered (by scenario modelling) to reduce the risk of resistance to combined *Bt* toxins by about ten-fold.[32] It is also likely that Bollgard II cotton coupled with HT traits will see GM cotton varieties become an important cornerstone of sustainable cotton production in Australia and elsewhere.

The International Cotton advisory Committee report[32] states that *Bt* cotton is likely to substantially improve economic benefits in developing countries, where pests substantially reduce yields despite use of pesticides (*e.g.* India). Qaim and Zilberman[33] argue that gains in realised cotton yields from growing *Bt* cotton will be most significant in South Asia and sub-Saharan Africa, with the added benefits of reducing poisonings to humans and farm animals arising from poor pesticide practices. Experiences of pesticide use on conventional cotton from small farms in mainland

China, South Africa and Mexico indicate potential advantages of cultivating *Bt* cotton in future.[32] The claimed advantages include increased income levels to resource-poor farmers, with significant flow-on gains for communities. A more cautious view on the value of *Bt* cotton for sub-Saharan African farmers is presented by de Grassie.[34] He claims that poverty in this area is not caused by poor cotton production technology and that *Bt* cotton has caused prices to fall in South Africa by 40%, with more than 60,000 farm workers in the cotton sector losing their jobs. He refutes the claims that the use of *Bt* cotton in South Africa has greatly reduced pesticide sprays and says that IPM measures have not been explored to their full extent. A lack of institutional capacity-building is blamed for the failure to reform the 'disconnected and top-down' system of agricultural research and development.

The potential risks and benefits of pest-resistant GM crops in Central and South America have remained a controversial issue since the reports that the GM trait for producing *Bt* toxin has been detected in Mexican landraces of maize and wild relatives (teosinte), presumably introgressed from *Bt* maize grown illegally in Southern Mexico.[35,36] Ongoing studies should reveal whether single gene traits (*e.g.* introgressed genes for expression of *Bt* toxins in teosinte maixe) will reduce or displace genetic diversity in important centres of genetic diversity for crops. These longer-term studies are based on monitoring the outcome of the selection pressure for increased, neutral or reduced ecological fitness when the hybrid plant is faced with constantly changing selection pressures from abiotic and biotic factors. CIMMYT's view[37] is that Mexican landraces are constantly evolving, while local farmers maintain the traits they desire. Whether this strategy maintains sufficient biodiversity for future generations of maize breeders and farmers remains to be seen. Hruska[38] cited several potential advantages and disadvantages for the use of GM crops in Central American agriculture. Reduction in pesticides (*e.g.* from cultivation of *Bt* cotton) could reduce production and human health costs, as well as benefiting the environment. He views GM crops as being easier to use (technology is 'all in the seed') than current IPM systems. However, this overlooks the considerable training required to manage *Bt* crops effectively, as demonstrated in Australia. Other potential disadvantages of *Bt* crops in Central America[38] include reduced consumer acceptance, affordability of seeds for resource-poor farmers, gene flow and genetic erosion of native wild relatives/landraces, development of *Bt* toxin resistance in pest populations, adaptation of GM crops to local conditions and intellectual property rights. The author advocates a pro-active approach to stimulate use of the right genes for the right reasons, and in the right way, preferably developed locally by regional research institutes or NGOs rather than by private, multi-national companies.

Below-ground Interactions with Agro-ecosystems, Using Bt *Crops as an Example*

Soil Eco-systems, Biodiversity and Function in Relation to GM Crops Soils contain the most diverse of eco-systems with many thousands of different species of bacteria, protozoa, fungi, micro- and macro-fauna. Numbers and activities are both temporally and spatially very variable. The bacterial and fungal communities perform many functions and transformations, such as transformations of mineral nitrogen for plant growth, plant growth promotion, pathogen inhibition and phosphorus mobilisation.[39]

Soil macro-organisms, the earthworms, nematodes and arthropods, feed on living and dead plant tissues, breaking the plant material into smaller pieces, and redistributing it so that it is more readily available for microbial activity. These chemical and physical processes are vital to biosphere functioning, providing the resources for continued plant growth and so the maintenance of all terrestrial ecosystems.

This wide diversity of the soil biota results in extremely complex food webs that are subject to a wide range of interactive influences. Soil environments are highly heterogeneous, and physical conditions and chemical gradients change spatially and temporally. As a result functional dynamics of the wide variety of biogeochemical processes that occur are also spatially and temporally very variable. Consequently species and functional process diversity in plant-soil systems is immense, producing a vast range of compounds that may be further transformed in other processes. These spatial and temporal variations over small to large scales make system predictions extremely difficult.

Activity in the soil ecosystem is normally limited by the energy availability, from fixed carbon compounds derived from plant primary production. These include inputs from plant roots, such as root exudates, cellular remains, root debris, and plant residues that fall to the soil surface (leaves, stems, flowers and fruit). These compounds are the source of both energy and nutrients for the soil biota. Although biological activity can occur in the bulk soil, most occurs close to the roots, in the rhizosphere, where plant inputs are the greatest. These inputs differ with both the type of plant and the growth stage. So plant species and genotype, and soil physical conditions, determine the amounts and types of compounds entering the soil, with consequential effects on microbial and macro faunal functional and population dynamics.[40,41,42] Soil ecosystem functional dynamics are also affected by other factors such as weather and cultivation, but soil fertility is dependent primarily on microbial activity, which is in turn responsive to plant inputs.

As well as providing an energy source, these carbon inputs can also affect microbial functioning in more subtle ways, as they are involved in microbial interactions and signalling. Temporally variable interactions between general soil heterotrophs and specific groups of micro-organisms, such as the autotrophic nitrifiers, occur in arable soils in response to additions of carbon and nitrogen. Microbial processes have been shown to be particularly responsive to protein substrates and carbon to nitrogen ratios.[43,44] Transgenic insecticidal plants such as *Bt* cotton produce and release relatively large amounts of a variety of novel proteins, many of which are active toxins.

Because of potential pleiotropic effects of the GM transgene and other changes arising from the GM crop transformation process itself, transforming a plant can have consequences, other than that specifically designed, on its physiology. Studies on transgenic plant residues in soil have shown differences in the nutritive quality of the transgenic plant for soil microbial communities and macro-invertebrates compared both to the isoline and to other cultivars.[14,45–48] Therefore the nutritive status of the plant will change, which will have consequential effects on soil functions such as degradation dynamics. This also means that the non-transformed parental or near-isoline plants may not provide completely satisfactory controls.[18] Any comparative experiments may not be only assessing effects of the toxins, but also changes in the organic inputs resulting from these differing plant physiologies; hence the requirement to compare the respective nutritive qualities of the transgenic and control plants prior to any functional determinations, such as degradation (*e.g.* cellulose and lignin content, amino acid content and ratios, C:N ratio, *etc.*).

Another factor requires consideration when *Bt* transformed plants are cultivated. *Bacillus thuringiensis* bacteria produce protoxins that require activation in specific gut sites in the target insects. In contrast, most GM *Bt* plants produce truncated and activated *Bt* toxins. Since these toxins do not require activation in the gut of a sensitive insect there is a possibility that other non-target macro faunal and microbial members of the soil ecosystem may also be affected *via* mechanisms which are not yet understood but are under study currently.

Input Routes of Bt *Plant Material and* Bt *Toxin into the Soil* Active *Bt* toxins are produced in all the cells of *Bt*-transformed plants, and these toxins can enter the soil eco-system by various routes.[49] There will be a continuous direct input *via* root exudates that increases in both quantity and spatial influence as the root system grows. *Bt* proteins can also be introduced through sloughed-off root debris, *e.g.* root cap cells and root hairs, again a continuously increasing input during the whole growing season. Fuchs[50] reported the occurrence of the *Cry1Ac* protein in roots of cotton with Event 531 (Ingard cotton). Although Saxena *et al.*[51] reported that *Bt* cotton does not exude the *Bt* proteins in the root exudates, unlike *Bt* maize, potato and rice, *Cry1Ac* has been reported at concentrations of between 1 and 43 $\mu g\ g^{-1}$ dry weight of roots four to nine weeks after germination, and was released from the roots into soil during growth.[52] Root breakage significantly increased *Cry1Ac* release into the soil ecosystem. As most of the activity in the plant-soil system occurs in the rhizosphere and is driven by the plant inputs, the possibility that the *Bt* proteins may have a direct effect on all the biotic dynamics associated with plant production must be investigated.

Input of the *Bt* toxins from the aerial parts of the plants into the soil occurs in two different ways. There will be a continual input during the whole growing season, from leaves, flowers and pollen, *etc.* falling to the ground. This will also increase as the plants develop during the growing season. The second is an annual, or biannual, input of relatively large amounts of plant residues, both dead and alive, *e.g.* stalks and seeds, left after harvest, particularly during cultivation for the following crop.

Sources of *Bt* protein input and the pathways and processes by which these affect soil eco-systems are summarised in Table 1. In this table, any item in the source column can be linked *via* any of the properties in the pathways and process columns to result in any of the outcomes in the effects column, *e.g.* pollen (source) may be ingested by fauna (pathway), degraded (process), and so decayed (effects).

There are several reports of the long-term persistence of active *Bt* proteins in soils in which *Bt*-transformed crops have grown.

Bt *Protein Persistence in the Soil Ecosystem* *Bt* toxins from *Bt* cotton plant material, with insecticidal activity, have been reported in soil after 28 days,[53,54] 120 days[55] and 140 days.[53] Greater persistence, of 234 days, has been reported for microbially produced *Bt* toxins[56] and from *Bt* maize residues.[45,49,57] *Bt* proteins can therefore

Table 1 *Pathways of GM crop inputs into soil ecosystems and possible effects*

Sources	*Pathways*	*Processes*	*Effects*
plant residues exudates pollen DNA transfer (plant/macro- micro-biota)	soil protein decomposition faunal ingestion ingestion transfer	adsorption denaturation degradation plant uptake elimination leaching run-off	decay rate persistence bioactivity accumulation

persist in soil for long enough to possibly affect following crops in the rotation. However, no detectable *Cry1Ac* toxin was found in soil samples three months after post-harvest tillage in a six-field study. *Bt* cotton had been grown in these fields for three to six years previously and the crop residues incorporated into the soil after harvest.[58] Three of the fields were sandy loam and three silt loam, with no reported differences between soil types.

Bt toxins are bound to clays and humic acids within a few hours of entering the soil, and can then remain bound for long periods.[59–62] Contrastingly, 10 to 30% of the *Bt* toxins can be leached from soils with low organic matter and high sand contents.[53,57] Adsorption is optimum at pH 6. So soil type is important to toxin persistence, with higher clay contents resulting in greater persistence. Over 80% of the micro-organisms in soil are adsorbed onto organic matter and clay minerals[63] and therefore in close proximity to any of the adsorbed *Bt* proteins. These active toxins may have an enlarged target range and so possibly adversely affect both the microbial and macro-faunal components of the eco-system, with consequences for functional dynamics.

Toxin Uptake by Plants, Soil-dwelling Micro-organisms and Macro-fauna As roots can re-absorb previously exuded organic compounds, a means of uptake of insoluble cations, it is possible that non-transformed plants may absorb *Bt* proteins using the same mechanism. However, when maize, carrot, radish and turnip plants were grown in soil that had previously grown a crop of *Bt* maize, or to which *Bt* maize residue or purified *Bt* toxin had been added, no toxins were found in any of them after 120 or 180 days.[64] Soil fauna will ingest *Bt* toxins when they feed on *Bt* plant roots and residues and absorb *Bt* proteins bound to humic acids and clays by physical contact. The *Bt* proteins themselves may also be a food source for soil herbivores, which may then sequester them and pass them up the food chain. Trophic relationships in soil are very complex. Yu *et al.*[65] reported no detectable effects of *Bt* cotton leaf tissue on two detritivores, the springtail *Folsomia candida* and the mite *Oppia nitens*. However, as these species are fungivores, *Bt* material may not have been directly consumed, as there are no reports that fungi growing on decaying *Bt* plant material contain *Bt* toxins, *i.e.* the exposure route requires further research.

Toxin Inactivation or Microbial or Chemical Degradation There are several reports that *Bt* proteins can be rapidly degraded microbially,[53–55,57,66] as can *Bt* maize residues.[49,67,68] Concentrations of the soluble toxins decline rapidly, followed by a more gradual decline to low concentrations that remain almost unchanged for several weeks or months.[49,53] However, any *Bt* proteins bound to clay minerals or humic acids in the soil are resistant to microbial degradation,[62,69] but *Bt* toxins in plant residues on the soil surface can be deactivated by sunlight.[69]

Horizontal Gene Transfer in Soil Ecosystems Nielsen *et al.*[70] reviewed the possibilities of horizontal gene transfer in the rhizosphere of transgenic plants. For natural transformations to occur in soils, free DNA and competent bacteria have to be in close proximity.[71] This could occur close to the roots when active degradation of *Bt* toxin-containing residues is occurring. Marker genes from some transformed plants have been detected in soil. Widmer *et al.*[72] reported that marker genes from tobacco and potato were still detectable at 77 and 137 days after the crop. In another study DNA of transgenic sugar beet plants was detectable for several months in the field.[73] Despite this evidence of persistence of plant DNA in soil, there are no reports of the transformation of plant DNA to indigenous soil micro-organisms. However, studies using sterile soil inoculated with a naturally transformed bacterium of *Acinetobacter* sp. showed recombination with transgenic plant DNA fragments.[73,74] It appears that the factor limiting horizontal gene transfer is the availability of competent cells in close proximity to any transformable DNA. Non-competent *Acinetobacter* sp. cells have been stimulated to become competent by a variety of inorganic salts and simple carbon sources that can be found in root exudates.[70] So it seems possible that transfer of genes from plants to bacteria may possibly occur, although this would only be in very restricted sites. For any subsequent exposure analysis all input and exposure routes by which soil biota can be exposed to transgenic plant material and the *Bt* toxins must be identified. Then soil functions can be ranked to identify which are of the highest priority for any pre-release impact assessment of transgenic plants on specific soil-ecosystems.

Possible Effects on Soil Ecosystem Functional Dynamics Soil ecosystems functional dynamics are dependent on the breakdown of plant residues to provide energy for a huge range of soil organisms. The quality of this input determines the dynamics of microbial and macro-faunal function. Coincidentally to this breakdown of plant residues, for energy, nitrogen-containing compounds required for many other microbial functions and continued plant growth are released. So plant–microbe relations have mutually beneficial consequences. As the constituents of such inputs, types of compounds and ratios, are crop-specific, it can be anticipated that microbial dynamics under transgenic plants will be directly affected by these plants. This is so whenever the crop plant is changed; therefore, any assessments of the effects of the GM transgene need to be designed to allow for this. Moreover, such changes are not irreversible, as subsequent dynamics will be dependent on the plants growing at that specific time. Although such changes in microbial populations may be described as 'transient', the possibility that repeated cultivation of *Bt*-transformed plants over many seasons cultivation may result in such 'transient' changes becoming permanent requires investigation.

Prioritisation of Soil Ecosystem Functions As soil-ecosystems are extremely complex, it is impractical and unreliable to study them on a species basis. There are less functional properties, but even so it is still impractical to measure them all, so choices of reliable parameters are required. One approach to this requires that a list of regionally and ecologically appropriate soil functions can be compiled, together with any associated soil biota, and then assessed using a selection matrix.[12] Soil functions can be divided into five main categories for further consideration: (a) degradation of the *Bt* protein-containing plant residues, (b) biogeochemical cycling, (c) plant/micro-organism/macro-faunal interactions, (d) crop pests and diseases, and (e) the role of biological activity in soil chemical and physical properties. The maximum potential significance of any adverse effects on these functions can be ranked on a 1, 2, 3 basis, ranking 1 having the greatest priority and 3 the lowest, for consideration of the importance of the function as an indicator of soil health, and how a variation in that function may affect crop development. Functions with the highest priority are important as indicators of soil health, so any adverse effects on functions are likely have a direct impact on crop development and yield.

For example, changes in the dynamics of soil organic matter decomposition will have consequences such as changes in the energy fluxes and nutrient supplies to other microbial processes. These will ultimately affect plant growth and soil aggregate stability, with consequent effects on root development and holding capacities, so this is ranked 1. As nitrogen-cycling dynamics in agricultural environments are known to be entirely dependent on the quality and quantity of plant inputs, the functions might be affected by input changes, particularly if there are any accompanying potential toxicity effects, and so the two important steps in nitrogen cycling in crop production (ammonification and nitrification) are also ranked 1. A possible impact on disease transmission might have significant consequences and, although it might be considered to be likely to appear over a relatively longer timescale, can similarly be ranked 1.

Due to their complexity, macro-invertebrates are best assessed as functional assemblages.[75] Using a multi-taxa approach removes any problems caused by a lack of detailed species-specific information. The macro-fauna can then be divided into three cross-taxa functional groups: (a) decomposers, (b) root feeders, and (c) disseminators. A major concern is that adverse affects on their functioning will reduce both residue breakdown rates and incorporation into soil organic matter, with consequential effects on the rate of organic matter decomposition by the microbial community. So the macro-invertebrate disseminator species are also ranked 1.

The recycling of inorganic nitrogen from plant residues for further crop production is a vital function of the soil ecosystem. A vast array of micro-flora and higher trophic groups of organisms, such as the micro- and meso-fauna, are interactive in this first step in the nitrogen cycle, ammonification. Ammonifying and nitrifying bacteria will come into direct contact with the *Bt* toxin in the rhizosphere and around the decomposing plant residues. The toxin might affect their activity and so the rate of ammonification and nitrification in the rhizosphere, decreasing the availability of nitrogen to the plant, and also consequently changing the rate of nitrification. Previous work[76,77] has shown that nitrification rates are particularly susceptible to changes in carbon inputs, particularly proteins.

Impacts of Bt *Toxins and Transformed Crops on Soil Eco-systems: Microorganisms*
Studies on different transgenic crops have found differences in microbial and fungal

community structure in transgenic plant rhizospheres, compared to the non-transgenic control.[78] These studies used molecular techniques that describe microbial populations as a whole or in constituent parts, such as the fungi, or functional groups, such as the nitrifiers, dependent on the primers used. Variable regions of 16S ribosomal genes are amplified from soil DNA extracts by PCR, using specifically targeted primers. The products can then be separated on a gel by differential gradient gel electrophoresis (DGGE) or temperature gradient gel electrophoresis (TGGE).[79] A similar technique is terminal restriction fragment length polymorphism (T-RFLP).[80–82] Additional information can be obtained from community-level physiological profiling (CLPP) using Biolog plates[83,84] and from phospholipids fatty acid profiles (PLFA).[85]

Such methods overcome the limitations inherent in using cultural techniques for identification and to describe populations. The main concern with these is that only a small fraction, which may not necessarily be the most common components, of the microbial community can successfully be cultured in the laboratory.[86] Investigations of whole microbial population DNA by DGGE have shown changes in the whole soil population profile during the crop-growing season and under different cultivated crops.[71,87] The methods are not quantitative, but the presence or absence of bands shows an appearance or disappearance of a group, and differences in the intensity of bands indicate that bacterial numbers are changing, even if the group is still present. The banding patterns can be compared between the transgenic plant and non-transgenic controls, taking care to compare the same field sites and stages of crop growth.

There are several reports of differences in the microbial populations associated with transgenic plants. Donegan *et al.*[48] reported a transient increase in fungal and bacterial populations when cultured on *Bt* cotton leaves. Differences in the carbon content of the transgenic and parental material were noted in a later experiment with buried litter bags,[48] together with differences in nematode and Collembola numbers in the surrounding soil. A study of any possible effects of *Bt*-transformed canola on soil populations showed that, although populations of bacteriophagous nematodes were no different, fungal feeders were more abundant compared to isogenic (control line) canola.[88]

Changes in the microbial communities may have adverse effects on functional dynamics. The microbial community is a main food source for many of the soil macro-fauna and so soil macro-invertebrates may also be adversely affected if microbial diversity is reduced. It is important to consider that microbial communities with similar structures as determined by these methods may still have ecologically significant differences in species composition, as the method is not sensitive to changes in community structure that may occur at the level of individual strains or species. The methods only assess changes in the numerically dominant populations of bacteria in a soil. Rare microbial populations are not represented because the template DNAs from these populations represent a small fraction of the total community and are not amplified or are present at levels that are not detected above the background.[80] Therefore these methods still only provide a limited answer to a specific hazard hypothesis. But they do indicate changes in community structure, through a comparison of gels and principal components analysis of the data, which may or may not have consequences for soil eco-system functioning.

Impacts of Bt *Toxins and Transformed Crops on Soil Eco-systems: Macro-organisms*
A study of the impact of leaf material from three *Bt* cotton lines found no differences in the numbers of protozoa, but did find changes in culturable bacterial diversity, and significantly greater increases in culturable bacterial and fungal population levels with the transgenic material, compared to the parental line, in the two weeks after the start of the experiment.[54] At the end of the experiments (28 or 56 days) these changes were no longer observed, suggesting that the transgenic plants may decompose faster than the parental plants. On two sampling occasions there was significantly greater utilisation of asparagine, aspartic acid and glutamic acid in soil with material from the two transgenic lines compared to the parental line. These substrates are important intermediates in nitrogen assimilation reactions. Because the changes were only observed for two of the transgenic lines and not the third, and not for the purified toxin, the authors conclude that they might be due to unintended changes in those transgenic lines rather than directly due to the *Bt* toxin. Studies on other transgenic plants also have found differences in microbial communities associated with the plants at the senescence growth stage, indicating an association with the decomposer community.[82,89,90]

Earthworms are a suitable indicator for the functional group of residue disseminators to examine any possible effects of the transgene on macro-faunal activity. Zwahlen *et al.*[91] reported no lethal effects of *Bt* maize litter on immature and mature *Lumbricus terrestris*, but reported a slight but significant weight loss after 240 days exposure, compared to worms eating litter of the non-transformed isoline. However, the earthworms did ingest the *Bt* toxin and excrete it in a concentrated form in their casts. Casts from *Bt* maize-fed earthworms were found to be toxic to the lepidopteran tobacco hookworm (*Manduca sexta*).[45] Earthworms may be sub-lethally affected by *Bt* proteins in the residues and soil they consume and, consequently, incorporation and dissemination of plant material in the soil might be affected. This may reduce the rate of decomposition of organic matter in the soil of *Bt*-planted fields. There may also be tri-trophic effects on their predators from *Bt* in their guts and on detritivores *via* the *Bt* toxin in their casts.

Residue-eating macro-organisms may also be affected, as the nutritional quality of the transgenic plants will differ from the controls. As well as any immediately lethal effects on macro-faunal members of the soil community, long-term sub-lethal and nutritional effects need to be considered. Any changes in macro-faunal activity and interactions may result in decreased plant residue diminution and incorporation rates. Currently there are few publications in this area but several EU (*e.g.* 'ECO-GEN') and international projects (*e.g.* GMO Guidelines projects) are starting to report results from multi-season studies on soil ecology interactions of GM crops.

3 Discussion of GM Crop Impacts

From our review and current literature on GM crop impacts it is clear that *Bt* crops, like all 'new' crop protection technologies in agriculture, offer potential benefits to growers. However, a range of case-specific potential risks also are identified, which vary according to the crop, the GM event (gene product, expression level, tissues expressed, genetic background of the variety, *etc.*), the receiving agro-environment (regional unless countries are uniform) and the management practices developed for safe use. The challenge for scientists and regulators is to respond objectively to a diverse range of sometimes conflicting stakeholder concerns. Farmers want higher and more stable yields but produced with reduced inputs of pesticides and fertilisers. Consumers want better quality foods, reduced pesticide residues and a cleaner environment. Policy makers want improved food security, decreased energy demands and greenhouse gas emissions, decreased environmental harm and want sustainable crop production methods to be increasingly used. The GM crop issue is clearly much wider than just GM crops, since it is inter-woven with wider issues about present-day agriculture, food production, environmental stewardship, international trade and global poverty. Currently consumer demand for GM crops and food is weaker in Europe than in the current main producing countries like USA, Canada, Argentina and China. Many consumers worried by media headlines and adverse publicity differentiate between GM crop types and seem more likely to accept non-food GM crops (*e.g. Bt* cotton) than food GM crops like maize, soybean and rice. This is possibly because of serious limitations in our risk foresighting (*e.g.* mad cow disease, foot and mouth disease, food contamination with Sudan I dye *etc.*) and public distrust of the key players involved (particularly large biotechnology companies and politicians involved in World Trade Organisation (WTO) negotiations on global trade issues). Whether we are talking about free trade at the level of the European Community, or the North American Free Trade Agreement, or the WTO, or individual developing countries, the rules inevitably mean that local (regional or country) values can come into conflict with the aims of global economic liberalisation.[92]

Even after ten years of growing GM crops around the world there is still a strong need for independent research and clearer guidelines for risk assessment testing, post-release monitoring and regulation. Future benefits and risks of current and next generation GM products are often incalculable or difficult to quantify in particular environmental and economic settings. This uncertainty about GM crops has contributed to a scientific and regulatory divide between those that follow 'precautionary approaches' and those that argue that if the evidence of adverse effects is small or incomplete (often from short-term or small-scale, studies or geographically dissimilar countries) then we should grow GM crops more extensively and rely on post-release monitoring to detect any future environmental problems. In a recent (2005) consultation of FAO experts[93] on monitoring environmental effects of GM crops it was recommended that any responsible deployment needs to address the whole process of technology development, from pre-release risk assessment to biosafety considerations and post-release monitoring. Potential hazards associated with GM cropping (the GM crop and its management systems) have to be placed within the broader context of both positive and negative impacts that are associated with all agricultural practices.[93] These experts agreed that consumers, farmer groups, environmental organisations and community groups need to be engaged in the scientific process of determining the local risks and benefits of each new GM crop event in heterogeneous farming systems. In summary, we agree with the FAO Director-General, Ms Louise Fresco, who recently stated that 'the need to monitor both the benefits and potential hazards of released GM crops is becoming ever more important, with the dramatic increase in the range and scale of their commercial cultivation, especially in developing countries'. International scientists are now helping developing countries to carry out their own risk assessments on regionally appropriate GM crops[94] so that particular GM crop types or varieties can be assessed and optimised for particular agricultural problems, local socio-economic needs and regional or country-specific environmental concerns.

Acknowledgements

The authors wish to acknowledge funding and support from Scottish Executive Environment and Rural Affairs Department (SEERAD) and from the Swiss Agency for Development and Cooperation (SDC).

References

1. S.R. Palumbi, *Science*, 2001, **293,** 786.
2. G.P. Georgiou, *Pesticide resistance: strategies and tactics for management*, National Academy Press, Washington, USA, 1986.
3. G. Lovei, *New Zealand Crop Prot.*, 2001, **54**, 93.
4. R. Costanza, R.R. d'Arge, R. De Groot, S. Farber, M. Grasso, B. Hannon, S. Naeem, K. Limburg, J. Paruelo, R. O'Neil, R. Raskin, P. Sutton and M. van den Belt, *Nature*, 1997, **387**, 253.
5. Natural Research Council, *The future role of pesticides in US agriculture*, National Academies Press, Washington, DC, 2000.
6. F. Gould, *Ann. Rev. Ent.*, 1998, **43**, 701.
7. A.N.E. Birch, B. Fenton, G. Malloch, A.T. Jones, M.S. Phillips, B. Harrower, J.A.T. Woodford and M.A. Catley, *Insect Mol. Biol.,* 1994, **3**, 239.
8. A.T. Jones, W.J. McGavin and A.N.E. Birch, *Ann. Appl. Biol.,* 2000, **136**, 107.
9. C. James, *Preview: Global status of commercialized transgenic crops*, International Service for Acquisition of Agri-Biotech Applications, 2004, 1.
10. F. Wu and W. Butz, *The future of genetically modified crops: Lessons from the Green Revolution*, 2004, *http://www.rand.org.publications/MG/MG161.*
11. G.M. Poppy and J.P. Sutherland, *Physiol. Entomol.*, 2004, **29**, 257.
12. A.N.E. Birch, R. Wheatley, B. Anyango, S. Arpaia, B. Capalalbo, E. Getu Degag, E. Fontes, P. Kalama, E. Lelmen, G. Lovei, I. Melo, F. Muyekho, A. Ngi-Song, D. Ochieno, J. Ogwang, R. Pitlelli, T Schuler, M. Setamou, S. Sithananthan, J. Smith, N. Van Son, J. Songa, E. Sujii, T. Tan, F-H. Wan and A. Hilbeck, Biodiversity and non-target impacts: a case study of Bt maize in Kenya, in *Environmental Risk Assessment of Genetically Modified Crops*, Vol. 1, A. Hilbeck and D.A. Andow (eds.), CABI International, Wallingford, 2004, 117.
13. A.N.E. Birch, I.E. Geoghegan, D.W. Griffiths and J.W. McNicol, *Ann. Appl. Biol.*, 2002, **140**, 143.
14. S. Flores, D. Saxena and G. Stotzky, *Soil Biol. Biochem.,* 2005, in press.
15. A.N.E. Birch, I.E. Geoghegan, M.E.N. Majerus, C.A. Hackett and J. Allen, *Scottish Crop Research Institute Annual Report 1996/7*, 69.
16. A.N.E. Birch, I.E. Geoghegan, M.E.N. Majerus, J.W. McNicol, C.A. Hackett, A.M.R. Gatehouse and J.A. Gatehouse, *Mol. Breed.,* 1999, **5**, 75.

17. A. Groot and M. Dicke, *Plant J.*, 2002, **31**, 387.
18. D.A. Andow and A. Hilbeck, *Bioscience*, 2004, **54**, 637.
19. G.L. Lovei and S. Arpaia, *Ent. Expt. Appl.*, 2005, in press.
20. A. Dutton, J. Romeis and F. Bigler, *Biocontrol*, 2003, **48**, 611.
21. W. Freese and D. Schubert, *Biotechnology and Genetic Engineering Reviews*, 2004, **21**, 1.
22. E.J. Calabrese and L.A. Baldwin, *Nature*, 2003, **421**, 691.
23. A. Wilson, J. Latham and R. Steinbrecher, *Genome Scrambling – myth or reality?* 2004. Free online download from *http://www.econexus.info.*
24. A.G. Haslberger, *Nature Biotech.,* 2003, **21**, 739.
25. G.G. Lovei, F. Szentkiralyi and S. Arpaia, *Ent. Expt. Appl.*, 2005, in press.
26. D.A. Andow, *Environ. Biosafety Res.*, 2003, **2**, 1.
27. D.N. Alstad and D.A. Andow, *Agbiotech News and Info.*, 1996, **8**, 177.
28. B.E. Tabashnik, Y. Carriere, T.J. Denney, S. Morin, M. Sisterton, R.T. Roush, A.M. Shelton and J-Z Zhao, *Insect resistance to Bt crops: Lessons learnt from the first seven years*, ISB News Report, November 2003, 1.
29. B.E. Tabashnik, Y. Cariere, T.J. Dennehy, S. Morin, M.S. Sisterton, R.T. Roush, A.M. Shelton and J-Z Zhao, *J. Econ. Entomol.*, 2003, **96**, 1031.
30. J. Fernandez-Cornejo and W.D. McBride, *Adoption of Bioengineered Crops*, Agricultural Economic Report No. 810, Economic Research Service, USDA, 2002.
31. G.P Fitt, Implementation and impact of transgenic Bt cottons in Australia, *Proceedings of the Third World Cotton Research Conference, Pretoria, South Africa*, 2003, 371.
32. G.P. Fitt, P.J. Wakelyn, J. Stewart, D. Roupakias, K. Hake, J. Pages and M. Giband, *Global status and impacts of biotech cotton*, Report on the Second Expert Panel for the International Cotton Advisory Committee, 2004, 1.
33. M. Quaim and D. Zilberman, *Science*, 2003, **299**, 900.
34. A. de Grassie, *Genetically Modified Crops and Sustainable Poverty Alleviation in Sub-Saharan Africa: An Assessment of Current Evidence*, Third World Network, Africa, 2003.
35. J.P.R. Martinez-Soriano and D.S. Leal-Klevezas, *Science*, 2002, **287**, 1399.
36. D. Quist and I.H. Chapela, *Nature*, 2004, **414**, 541.
37. M. Bellon, J. Berthaud, D. Hoisington, M. Iwanaga and S. Taba, *Are Mexico's indigenous maize varieties at risk?* Download from CIMMYT website, *http:// www.cimmyt.org/ whatisiscimmmyt/recent_ar/D_Sustain?mexico.htm, 2004.*
38. A.J. Hruska, *Biotech Dev. Monitor* 1996, **29**, 79.
39. J.D. Van Elsas, J.T. Trevors and E.M.H. Wellington, *Modern Soil Microbiology*, Marcel Dekker Inc., New York, USA, 1997.
40. K.E. Dunfield and J.J. Germida, *FEMS Microbiol. Ecol.,* 2001, **82**, 1.
41. S.D. Siciliano, C.M. Theoret, J.R. de Freitas, P.J. Hucl and J.J. Germida, *Can. J. Microbiol.,* 1998, **44**, 844.
42. S.J. Grayston, S. Wang, C.D. Campbell and A.C. Edwards, *Soil Biol. Biochem.,* 1998, **30**, 369.
43. R.E. Wheatley, C.A. Hackett, A. Bruce and A. Kundzewicz, *Int. Biodeterioration Degradation,* 1997, **39**, 199.

44. R.E. Wheatley, K. Ritz, D. Crabb and S. Caul, *Soil Biol. Biochem.*, 2001, **33**, 2135.
45. D. Saxena, and G. Stotzky, *Soil Biol. Biochem.*, 2001, **33**, 1225.
46. D. Saxena, and G. Stotzky, *American J. Bot.*, 2001, **88**, 1704.
47. N. Escher, B. Käch and W. Nentwig, *Basic Appl. Ecol.*, 2000, **1**, 161.
48. K.K. Donegan, R.J. Seidler, V.J. Fieland, D.L. Schaller, C.J. Palm, L.M. Ganio, D.M. Cardwell and Y. Steinberger, *J. Appl. Ecol.*, 1997, **34**, 767.
49. C. Zwahlen, A. Hilbeck, P. Gugerli, and W. Nentwig, *Mol. Biol.*, 2003, **12**, 765.
50. R.L. Fuchs, Report No. MSL-13315, Monsanto Company, St. Louis, MO, USA, in OGTR (2003) *Risk Assessment and Risk Management Plan Consultation Version: Commercial release of insecticidal (INGARD® event 531) cotton*, Office of the Gene Technology Regulator, Woden, Australia, 2003.
51. D. Saxena, C.N. Stewart, I. Altosaar, Q. Shu and G. Stotzky, *Plant Physiol. Biochem.*, 2004, **42**, 383.
52. V.V.S.R. Gupta, G.N. Roberts, S.M. Neate, S.G. McClure, P. Crisp and S.K. Watson in *Biotechnology of Bacillus thuringiensis and its environmental impact*, R.J. Akhurst, C.E. Beard and P.A. Hughes (eds.), PACSIRO Entomology, Canberra, Australia, 2002, 191.
53. C.J. Palm, D.L. Schaller, K.K. Donegan and R.J. Seidler, *Can. J. Microbiol.*, 1996, **42**, 1258.
54. K.K. Donegan, C.J. Palm, V.J. Fieland, L.A. Porteous, L.M. Ganio, D.L. Schaller, L.Q. Bucao and R.J. Seidler, *Appl. Soil Ecol.*, 1995, **2**, 111.
55. S.R. Sims and J.E. Ream, *J. Agric. Food Chem.*, 1997, **45**, 1502.
56. H. Tapp and G. Stotzky, *Soil Biol. Biochem.*, 1998, **30**, 471.
57. D. Saxena and G. Stotzky, *FEMS Microbiol. Ecol.*, 2000, **33**, 35.
58. G. Head, J.B. Subber, J.A. Watson, J.W. Martin and J. Duan, *Environ. Entomol.*, 2002, **31**, 30.
59. G. Venkateswerlu and G. Stotzky, *Current Microbiol.*, 1992, **25**, 225.
60. H. Tapp, L. Calamai and G. Stotzky, *Soil Biol. Biochem.*, 1994, **26**, 663.
61. H. Tapp and G. Stotzky, *Appl. Environ. Microbiol.*, 1995, **61**, 602.
62. C. Crecchio and G. Stotzky, *Soil Biol. Biochem.*, 1998, **30**, 463.
63. M. Bruinsma, J.A. van Veen and G.A. Kowalchuk, *Effects of genetically modified plants on soil ecosystems*, Concept report for the Committee on Genetic Modification (COGEM), Netherlands Institute of Ecology, Heteren, Netherlands, 2002.
64. D. Saxena and G. Stotzky, *Nature Biotech.*, 2001, **19**, 199.
65. L. Yu, R.E. Berry and B.A. Croft, *J. Econ. Entomol.* 1997, **90**, 113.
66. D. Saxena, S. Flores and G. Stotzky, *Nature* 1999, **402**, 480.
67. S.R. Sims and L.R. Holden, *Environ. Entomol.*, 1996, **25**, 659.
68. D. Saxena, S. Flores and G. Stotzky, *Soil Biol. Biochem.*, 2002, **34**, 133.
69. J. Koskella and G. Stotzky, *Appl. Environ. Microbiol.*, 1997, **63**, 3561.
70. K.M. Nielsen and J.D. van Elsas, *Soil Biol. Biochem.*, 2001, **33**, 345.
71. K. Smalla, G. Wieland, A. Buchner, A. Zock, J. Parzy, S. Kaiser, N. Roskot, H. Heuer and G. Berg, *Appl. Environ. Microbiol.*, 2001, **67**, 4742.
72. F. Widmer, R.J. Seidler, K.K. Donegan and G.L. Reed, *Mol. Ecol.*, 1997, **6**, 1.
73. F. Gebhard and K. Smalla, *FEMS Microbiol. Ecol.*, 1999, **28**, 261.

74. K.M. Nielsen, J.D. van Elsas and K. Smalla, 2000, *Appl. Environ. Microbiol.*, **66**, 1237.
75. L. Mendoca-Hagler, I. Soares de Melo, M.C.V. Inglis, B. Anyango, J.O. Siqueira, P.V. Toan, R.E. Wheatley, Non-target and biodiversity impacts in soil, in *Environmental Risk Assessment of Genetically Modified organisms: A Case Study of Bt Cotton in Brazil*, **Vol. 2**, A. Hilbeck and D.A. Andow (eds.), CABI International, Wallingford, 2005, in press.
76. R.E. Wheatley, K. Ritz, D. Crabb and S. Caul, *Soil Biol. Biochem.*, 2001, **33**, 2135.
77. T.A. Mendum and P. R. Hirsch, *Soil Biol. Biochem.*, 2002, **34**, 1479.
78. K.E. Dunfield and J. Germida, *J. Environ. Qual.*, 2004, **33**, 806.
79. G. Muyzer and K. Smalla, *Antonie van Leeuwenhoek*, 1998, **73**, 127.
80. W.T. Liu, T.L. Marsh, H. Cheng and L.J. Forney, *Appl. Environ. Microbiol.*, 1997, **63**, 4516.
81. C.B. Blackwood, T. Marsh, S.-H. Kim, and E.A. Paul, *Appl. Environ. Microbiol.*, 2003, **69**, 926.
82. T. Lukow, P.F. Dunfield and W. Liesack, *FEMS Microbiol. Ecol.*, 2000, **32**, 241.
83. B.S. Griffiths, I.E. Geoghegan and W.M. Robertson, *J. Appl. Ecol.*, 2000, **37**, 159.
84. J.S. Buyer, D.P. Roberts, P. Millner and E. Russek-Cohen, *J. Microbiol. Method*, 2001, **45**, 53.
85. C.B. Blackwood and J.S. Buyer, *J. Environ. Qual.*, 2004, **33**, 832.
86. P. Hugenholtz, B.M. Goebel and N.R. Pace, *J. Bacteriol.*, 1998, **180,** 4765.
87. T. Pennanen, S. Caul, T.J. Daniell, B.S. Griffiths, K. Ritz and R.E. Wheatley, *Soil Biol. Biochem.*, 2003, **36**, 841.
88. B. Manachini, S. Landi, M.C. Fiore, M. Festa and S. Arpaia, *IOBC wprs Bulletin*, 2004, **27**, 103.
89. J. Lottmann, H. Heuer, K. Smalla and G. Berg, *FEMS Microbiol. Ecol.*, 1999, **29**, 365.
90. J. Lottmann, H. Heuer, J. de Vries, A. Mahn, K. During, W. Wackernagel, K. Smalla and G. Berg, *FEMS Microbiol. Ecol.*, 2000, **33**, 41.
91. C. Zwahlen, A. Hilbeck, R. Howald and W. Nentwig, *Mol. Ecol.*, 2003, **12**, 1077.
92. P. Sands, *Lawless World: America and the Making and Breaking of Global Rules*, Penguin, Allen Lane Books, 2005.
93. L. Guarneri, *Monitoring the environmental effects of GM crops*, 2005, Free download from *http://www.fao.org/newsroom/en/news/2005/89259/index.html.*
94. A. Hilbeck and D.A. Andow, *Environmental Risk Assessment of Genetically Modified Organisms: A Case Study of Bt Maize in Kenya*, **Vol. 1**, CABI International, Wallingford, UK, 2004.

Sustainable Land Management: A Challenge for Modern Agriculture

DANIEL OSBORN

1 The Agricultural Origins of Sustainable Development

The amount of the Earth's surface that represents land we can use for any purpose, whether in a sustainable fashion or not, is only a small fraction of the total surface. The amount of the Earth's surface that is underwater in lakes, rivers and the seas and oceans is less accessible and less used. For the deep ocean, so little is known of it that it is currently the subject of international research initiatives aimed at finding out just what is there. It is the small fraction of the land surface that we can use that has to support most of humanity's needs. And, as the human population grows, so do the pressures on not only the land we use for agriculture but also the ecosystem services it delivers to us all.

We need to find ways of using natural resources like soil and water in a sustainable fashion. At present this may not be the case for either of these key agricultural resources. Without adequate amounts of good quality soil and water it becomes difficult to imagine just how we can supply enough food for a growing world population. Of course, the chemical and biotechnology industries might perform yet another technological revolution. But there is no certainty that they are planning for one at present when agricultural surpluses (at least in the developed economies) seem to be a greater policy issue than food shortages and when there remains considerable public resistance to the introduction of new technologies in certain major world markets.

In the past, the use of land to meet human needs for food, shelter, transport and materials has expanded in two ways. First, in terms of spatial extent and second, in terms of intensity. Technological progress has been the lynch pin of both types of expansion and, to a degree, spatial expansion and intensification have tended to occur in areas where incoming technologies could best be deployed from either an economic

Issues in Environmental Science and Technology, No. 21
Sustainability in Agriculture

or social perspective or, preferably, both. Of course, some experiences with new technologies, or old technologies introduced in new places, shows that this has not always been the case (*cf.* dustbowls of the mid 20th century; aspects of desertification and overgrazing now). This suggests technological fixes cannot be introduced into ecosystems that are not sufficiently resilient to bear the impacts of their deployment.

Until relatively recently in human history, the main drivers of land use were probably rather localised needs for food and shelter. Impacts on the environment were relatively small scale and often pretty much the direct result of actions by relatively few individuals trying to meet their own needs. In many ways, people had to work with nature rather than trying to find ways of substantially circumventing the natural constraints that would otherwise limit our aspirations for development and improved livelihoods. Even so, food production has probably always been a highly managed process. For example, plant and animal breeding is just one way that people have tried to improve their lot over a period of about 3000 years. The risks of not managing food production are too great for it to be anything else.

During the course of the 19th and 20th centuries, evidence began to accumulate that human activity could have unintended effects. This challenged some earlier ideas that people could not do anything that would damage the planet. Moreover, these effects could be seen at some distance from the sources, and had a duration and severity of effect that could not be ignored. For example, emissions from factories so badly damaged agricultural and other types of rural land in England in the 19th century that parts of north-west England were described as resembling the surface of the moon. This was clearly an effect on the sustainability of agriculture or options for the future that in modern parlance we would describe as unsustainable. Strict pollution control laws (the Alkali Acts of the 1860s on) were introduced to place some constraints on economic development. Negative effects of agriculture were less apparent at first. But evidence progressively mounted, especially after the Second World War, that showed, for example, that some organochlorine pesticides (DDT, dieldrin) could kill non-target organisms at a great distance from the place of application after moving through food chains. The subtle nature of some of the effects (a metabolite of DDT caused the egg-shells of certain birds to get thin and break in the nest) were significant in that they illustrated that even without killing organisms negative ecological outcomes could occur.

Thus, in the second half of the 20th century some books and many academic papers appeared raising the possibility that economic and social development was unlikely to be without substantial, if not intolerable, risks for organisms and ecosystems unless specific steps were taken to limit economic growth. Some of these important (if not seminal works) – for instance, *The Limits to Growth*, a report sponsored by a group of business leaders, and Rachel Carson's *Silent Spring* – are now seen as the early forebears of what came to be a series of intergovernmental conferences on sustainable development.[1]

Many sectors of the economy must have feared that the early findings really did mean an end to high rates of growth. But between the early books and the later summits came the Brundtland Commission's *Our Common Future* which first (and perhaps best) defined sustainable development as 'meeting the needs of the present without compromising the ability of future generations to meet their own needs'.[2] At last, here was a concept that suggested that solutions to the environment–development conflict could be found if local people could find appropriate technologies for their circumstances and if they were also given the wherewithal to innovate. Since then a range of economists [3] have worked on the links between environment and economics. An enormous amount of progress continues to be made and a range of views on how human activity can be made more sustainable have been produced by economists, social scientists, international conferences and environmental scientists. The debate is not, even yet, producing clear-cut prescriptions as to what does and does not constitute sustainability.

But one thing has become more certain: unfettered use of natural resources is not a sustainable option for managing the global economy of a planet on which six billion people live and where the land surface for growing crops is finite and is a resource under pressure from processes such as desertification and soil erosion. Regional water shortages add to these problems. Some may argue that this view is not necessarily the case as humanity is so ingenious that it has always come up with a technological fix to solve even the most pressing problems. The counter to that today is that the rate at which human activity is changing even the basic composition of the atmosphere is probably far too fast for ecosystem processes to accommodate without them undergoing changes that could well make life much less tolerable and which might, *in extremis*, foster an economic decline that was more than just a matter of international relativities.

2 Agriculture: A Drive Towards More Sustainable Land Management

A key aspect of modern agriculture is how it uses land and other natural resources intensively to produce high yields of crops and other products. These processes depend on the use of fertilisers and agro-chemicals to control pests and diseases. Modern agricultural practices have, in the relatively recent past, prioritised food production above environmental concerns and, as a result, adverse impacts on biodiversity and water quality have been commonplace. These adverse effects are very specific to particular countries and regions. This is partly because of differences in management practices and partly because of natural environmental heterogeneity. The importance attached to these adverse effects also varies from place to place because people's attitudes differ depending on their cultural background, educational and social status, and economic circumstances.

In recent years, it has been recognised that the effects on water quantity and quality, rates of soil erosion and biodiversity have been reaching levels where the costs of mitigation and remediation are sufficiently high to effect taxpayers' pockets (*e.g.* water bills can nowadays reflect the need to protect environmental resources). Concurrently, in many regions of the world, particularly perhaps in north-west Europe, there are moves to reform 'institutions' such as the Common Agricultural Policy (CAP) in a way that redresses the balance between agricultural yields and environmental goods and services. In doing this, it has been accepted that any rebalancing of priorities would only be possible if land managers and owners were able to make a continuing living from the land. In short, there is now a substantial move to encourage agriculture to use land and crop management practices that are more sustainable.

3 Making Agriculture More Sustainable

In the United Kingdom and in other parts of the European Union, initial attempts at making agriculture more sustainable were, perhaps almost inadvertently, driven by the need to reduce the level of agricultural surpluses (the surpluses were referred to as 'mountains' in the 1980s) to an acceptable level. There is, of course, a case to be made for maintaining a certain level of surpluses lest the earth's ecosystems are subject to a major planetary shock, *e.g.* supervolcano eruption, asteroid strike, disease outbreak affecting one of the major crops (*e.g.* wheat, rice, soya). But this may not be the main driver for maintaining stocks at a level that is a little ahead of demand.

Restrictions on production levels led to land being set aside. This created a potential land reserve that could be used to address biodiversity imbalances that had arisen from intensive agricultural practices (such as the decline in farmland birds in the UK that had happened despite the withdrawal of the more toxic and persistent pesticides). However, in the absence of prescriptions on how to manage set-aside land for biodiversity, initial benefits were limited. Another problem was the gulf between scientists engaged on environmental research and the farming sector. This made it difficult to transfer knowledge from scientists to land managers on the one hand, and, on the other, made it difficult for the scientists to appreciate the practical and economic factors that constrain the options available to farmers.

From the mid 1990s onwards, however, this gulf has been closed by a number of interdisciplinary projects which have had the effect of engaging the stakeholders and allowing them to frame the nature of scientific research designed to help make agriculture more sustainable and, perhaps, reverse biodiversity imbalances or declines. These projects, known as SAFFIE and BUZZ – see the Framed Environment Company's website at *http://www.f-e-c.co.uk/buzz.htm* for information on BUZZ – are now delivery prescriptions that are part of the reform delivery mechanism for the evolving priorities of the Common Agricultural Policy.

The initial results from these studies are interesting. It would appear that farmers may be able to make a living by concentrating efforts to manage resources for biodiversity on the less productive areas of the farm (or those that are more awkward to manage). Early results from these studies suggest it may be possible to reverse the decline in such important sustainability indicators as the presence of farmland birds and keystone species, such as bumblebees. Interestingly, pesticide use is permitted even on the areas being managed for biodiversity – but use is selective in terms of both the choice of chemical and target species and is carefully managed to increase 'yields' of desired non-crop plants. That this approach is different is exemplified by the lack of an everyday word for desired non-crop plants – all we have (I think!) is a word for undesirables ('weeds').

4 Principles for Making Agriculture Sustainable

In order to make land management more sustainable, key elements of the international sustainability agenda need to be taken on board by any economic or social group trying to reach the goal of sustainability. These principles can be extracted from the United Nations document *Plan of Implementation of the World Summit on*

Sustainable Development, Johannesburg, 2002 [1] and from related documents prepared under the auspices of the United Nations Economic Commission for Europe.[4] They are:

- Patterns of consumption and production need to be examined throughout their complete life-cycle in order to determine the balance between economic and social advantage and environmental impacts. Such studies have to determine the capacity of the environment to produce a range of ecosystem services from any given parcel of land. This means it is not adequate, for example, for agriculture just to consider the role of soil in food production or animal husbandry. It is also necessary, for instance, to consider the function of soil in flood defence where the capacity of soil to absorb water can be a key mitigating factor. Sustainable agriculture systems would avoid impairing the latter function of soil.
- Stakeholders need to be aware of the options for managing land in different ways. In considering options, the risks and opportunities associated with each option need to be taken into account. Awareness of options, risks and opportunities will help them make an informed choice as to which option is the most sustainable given their particular set of social, economic and environmental circumstances.
- Stakeholders need to be engaged not only in the decision-making process but also in formulating information gathering and research activity designed to help them make informed choices (*i.e.* choices made with knowledge of the outcomes of decisions). The process of stakeholder engagement should make it possible to resolve conflicts over natural resources and devise sustainable management plans that should secure the livelihoods of future generations and the interests of the environment.
- Decisions on sustainable land-management may need to be made at geographical scales that are above the traditional decision-making unit of the field, farm or estate. In other words, land owners and managers may need to make decisions at the scale of the river catchments or the landscape. This would entail levels of co-operation between stakeholders of a very different kind to that which society is used to. One key to success would be the provision of information on options for land management at such scales.
- It seems likely that to make such land management practices work, indicators of sustainability will need to be developed using multi-criteria techniques. For such indicators, or for key components of multi-criteria indicators, there would need to be thresholds of acceptability that were based on an understanding of ecological and environmental processes regulating biodiversity or setting the rates of biogeochemical or hydrological cycles. Thresholds would have to be set such that remedial action could be taken before any problems became serious enough to do harm or damage.

Although adopting these principles is challenging for all concerned, it may well be that agriculture is a good area in which to test these ideas and establish good working practices. This is because the agricultural sector has already dealt with environmental issues over a number of years. For example, in the area of pest and disease control a system of regulation of pesticide use has evolved with a strong scientific base. In many senses this system of regulation has the effect of balancing, through a network of committees, the need for adequate supplies of good quality food against concerns for human health and environmental integrity. Good quality risk assessment has been the key to success in this area of agriculture.

But, even here, landscape level management has yet to be tried. There are few, if any, easy to use tools to help make such systems work and a number of open questions. For example, how might land owners that lost out in a landscape management plan be compensated for opportunities forgone?

Some aspects of game theory might be useful here, although its most advanced forms tend to concern players that do not co-operate but are rational and survive by using strategies in a game they have learnt the rules of. However, a key factor in the successful working through of such 'games' is the provision of information to all the players.[5]

5 Science for Sustainable Agriculture

Despite the good model that pesticide use and regulation might provide, we still need a better scientific basis for developing more sustainable agricultural practices. One reason for this is because there are a number of areas of risk assessment that still need to be developed – including ways of accounting for indirect effects, such as risks to vertebrates whose food supplies decline following loss of seed-setting plants from farmland after repeated herbicide applications over large areas of land. Another reason for needing a better science base concerns the lack of well-tested approaches for dealing with complete systems of production and consumption of which agriculture is only one. The concept of environmental appraisal, developed within the UK Foresight programme, might help here. This is because Environmental Appraisal refers to 'the systematic analysis and evaluation of the environmental effects and implications of human activities' and is 'driven by the need to inform decision makers of the likely environmental consequences of an activity and to choose between competing options'. [6] The concept has not been extensively tested.

Even if risk assessment and environmental appraisal need further work, science can deal with some of the information requirements of both game theory and environmental appraisal as it is clear that to make agriculture more sustainable we need an accurate view of the interplay between natural resources, agriculture and the ecosystem services people need from air, soil and water.

One approach to doing this is the survey and monitoring activities typified by the UK Countryside Survey (see http://www.cs2000.org.uk/ for more information). This survey makes observations on the stocks and trends in environmental resources and attributes of over 500 1 km squares at periods of at least once per decade (since the late 1970s). The next survey will, as a minimum, assess environmental stock and change in rural landscapes in Great Britain in 2006–2007[7] and will be able to determine not only what changes are occurring in a large number of environmental attributes, but will also say why these are happening and determine the relative importance of the forces driving change. This sample-based survey, combined with satellite and aircraft censuses of the whole of the land surface of Great Britain, provides a basis for understanding the relationship between environmental resources and gradients of agricultural intensity.

Such activities and others involving volunteers such as the UK Phenology Network (see *http://www.phenology.org.uk*), the biodiversity recording schemes run by the Biological Records Centre (see *http://www.searchnbn.net*) and the British Trust for Ornithology (see *http://www.bto.org/survey/index.htm*) provide ways of understanding how a range of pressures arising from human activity can affect natural resources and the capacity of the environment to renew these. All these surveys and surveillance systems provide information that can be increasingly interpreted using modern computational approaches, such as visualisations.

Such survey and monitoring activities, combined with knowledge of environmental and ecological processes make it possible to devise new prescriptions for land management that should be more sustainable. This is especially the case if experiments into alternative land management options are designed to take into account all relevant factors that are involved in the agricultural system as a whole. The SAFFIE, BUZZ and the recently completed Farm Scale Evaluations (see *http://www.defra.gov.uk/environment/gm/fse/*) indicate ways in which science can inform the debate on the sustainable development of agriculture in practical ways.

The importance of monitoring the outcome of land management decisions cannot be stressed enough. In the past, we failed to see whether our choices about land management had the expected outcome. Regular monitoring would help ensure we were making the right decisions. To be of real use in sustainable land management, monitoring would need to provide an early warning of adverse change. This means using signals that provide land managers with both diagnostic and prognostic indicators. It will be no use having an indicator that is effectively an emergent property of a complex system. This is because such indicators may only respond when a major ecological change is happening, or worse still, after it has happened. Indicator systems are required that provide a warning of impending change, and, best of all, suggest why the change is occurring and thereby the solution to the problem – if there is one. This will probably mean developing metabonomic indicators of the health of ecosystems subject to what we consider to be sustainable land management practices. These types of indicator are already being developed and used in human medicine and are informing the debate on the relationships between human health and the environment.

References

1. WSSD, *Plan of Implementation of the World Summit on Sustainable Development, Johannesburg, 2002*, United Nations, New York (available at: *http://www.johannesburgsummit.org/html/documents/summit_docs/131302_wssd_report_reissued.pdf*).
2. WCED, *Our Common Future. World Commission on Environment and Development*, Oxford University Press, Oxford, 1987.
3. K. Turner, D. Pearce and I. Bateman, *Environmental Economics: An Elementary Introduction*, Pearson Higher Education, Harlow, 1993.

4. UNECE, *Declaration by the Environment Ministers of the Region of the United Nations Economic Commission for Europe (UNECE) at the 5th Ministerial Conference, Kiev 2003*, United Nations, Geneva (available at: *http://www.unece.org/env/documents/2003/kievconference/ece.cep.94.rev.1.e.pdf*).
5. E. Klarreich, *The Mathematics of Strategy* (available at *http://www.pnas.org/misc/classics5.shtml*).
6. DTI, *Towards More Sustainable Decisions*, Foresight Environmental Appraisal Task Force, Department of Trade and Industry, HMSO, London, 2001.
7. Defra, *Securing the Future–the UK Government's Sustainable Development Strategy*, HMSO, London, 2005 (available at *http://www.sustainable-development.gov.uk/publications/uk-strategy/uk-strategy-2005.htm*).

UK Environmental–Economic Consequences of Decoupled CAP Payments

IAN DICKIE AND ANNA SHIEL

1 Introduction

The implications of the production subsidies paid under the EU's Common Agricultural Policy (CAP) have long been a hot topic for environmentalists. In June 2003, an agreement was reached to reform the CAP. A 'decoupled' (independent of production) single farm payment for EU farmers would replace the old subsidy system. In addition, the receipt of subsidies was made subject to basic agricultural and environmental standards.

In environmental terms, decoupling may not appear to be a significant change, but in economic terms, decoupling represents a significant shift in the policy signals facing farmers. The purpose of this article is to examine the scale of this shift and its implications for environmental policies in the UK.

UK policies in all sectors are subjected to detailed consideration of their potential effects on the economy, including any economic sectors likely to be affected.[1] When considering the likely effects of environmental policy changes on agriculture, the UK Government therefore seeks to assess the likely costs to the sector. Work on the control of diffuse pollution from agriculture by Oxera Consulting Ltd (personal communication) examined the potential costs of using economic instruments to address this issue. It identified the increased gross margin on produce as a result of price support under the CAP, and the resulting higher abatement costs of dealing with pollution. The implication of this is that decoupling will reduce pollution abatement costs, making adjustments to environmental requirements less expensive for the agricultural sector in the future.

This represents a potentially significant benefit from decoupling. More affordable environmental policies will improve the prospects for a sustainable future for agriculture, where businesses remain economically viable while maintaining and enhancing the quality of the environment, and contributing to social objectives.

Issues in Environmental Science and Technology, No. 21
Sustainability in Agriculture

This article examines the relationship between subsidies and the environment (Section 2), with particular reference to the CAP and decoupling (Section 3). We present two simple case studies of the costs of extensification of agricultural activity (Section 4) and examine the policy implications of the impact of decoupling on this typical environmental response (Section 5).

2 Subsidies and the Environment

Defining Subsidies

During the 1990s, total subsidies were estimated to make up 3.8% of a global economy of £26 trillion. However, the precise definition of subsidies is not straightforward,[2] involving interventions that affect suppliers of goods and services by lowering the cost of production or raising the price received, compared to those in a market undistorted by government. Subsidies can be distinguished from transfer payments, in that they have the object of keeping consumer prices below the cost of production.[3]

Subsidies can be regarded as a financial measure, including financial transfers to producers, regardless of whether targeted on products or simply in the form of cash sums payable to producers.[4] Under this definition the direct payments currently made to producers under the CAP, agri-environment payments and the proposed decoupled payments, are all subsidies.

Other analyses go further, defining both environmental as well as financial subsidies.[5] Financial subsidies include non-recovery of public management costs, favourable tax treatment, direct contributions and lower than normal rates of return. These types of subsidy have occurred in the UK in recent years; for example, farmers have received favourable tax treatment (*e.g.* on Red Diesel).

Environmental subsidies are defined as the non-payment of environmental disruption costs by the entities causing the disruptions.[5] The failure of governments to internalise these environmental costs has been referred to as a covert subsidy. In the UK, these costs include the costs of diffuse pollution from agriculture, which have been identified elsewhere.[6] This article concentrates on direct financial subsidies in agriculture, which are the subject of the current CAP reforms.

The Impacts of Subsidies

Neo-classical economics suggests that input and output price subsidies will promote intensification of production.[7] Indeed, it is their intention to bring higher levels of activity than provided by the market.[4] This can have both positive and negative consequences.

Positive Consequences of Subsidies Where markets are providing less than the socially optimal level of activity, a subsidy may help increase activity to reach that level. Subsidies are deployed to shield sectors or products from international competition, assist sectors of strategic importance, and support employment. They have been shown to stimulate private R&D.[8] Positive effects from subsidies include helping the poor, encouraging technological development and paying for environment benefits.[2] Agricultural subsidies in the UK have had positive consequences for incomes in farming, and may have helped slow down the exit of businesses from the sector.

Markets fail to deliver sufficient levels of public goods.[a] However, subsidies can make free markets work better. Agricultural markets will provide less than the optimum level of the environmental public goods (positive externalities) of agriculture. For example, while consumers can purchase agricultural goods produced to higher environmental standards (*e.g.* organic food), the environmental benefits of organic production techniques are non-excludable. Therefore, consumers cannot prevent others free-riding on the price premium they pay for the environment benefit.

Free-riding is a market failure, so if subsidies to agriculture correct this market failure, and increase the provision of public goods, then there is a gross benefit to society. This benefit provides the justification for public support for agriculture according to the Policy Commission on the Future of Farming and Food. The Commission stated in its recommendations that 'public funds should be refocused on public goods, rather than subsidising overproduction...we want to see the (European) Community's budget for environmental programmes in the countryside substantially increased, helping to encourage best practice and pay for environmental benefits which the market will not provide'.

Negative Consequences of Subsidies The net benefits of subsidies have been called into question; substantial attention has been given in recent years to the role that financial subsidies play in environmental degradation. Subsidies tend to be environmentally damaging and do not contribute to sustainable development; as they encourage the activity that is subsidised, if these activities have negative environmental impacts, then the subsidy exacerbates those harmful impacts.[2]

So where financial subsidies are associated with environmental subsidies (negative externalities) the environmental impacts (damages) tend to be magnified.[5] If the damages outweigh the benefits to society, the subsidy is providing a perverse incentive, encouraging activity that is not in society's overall interests. The negative consequences of subsidies include:[2,4]

- Expense for governments, and the money used could be spent elsewhere;
- Sub-optimal use of society's resources, such as inefficiencies in production;
- Rent-seeking behaviour and difficulty in managing over-capitalised sectors;
- Over-production and hence many associated effects such as pollution;
- Excess production must be 'dumped' or disposed of.

An example of the negative effects associated with subsidies is that in 1993 Japan's subsidies to rice producers meant that its proportion of the world's rice insecticide use was ten times its share of world rice production.[7]

[a] Public goods are defined as commodities, the consumption of which must be decided by society as a whole, due to their properties as non-excludable, non-rivalrous and non-rejectable.[3]

Tackling the Effects of Subsidies

The overwhelming view of subsidies is that they are not beneficial.[2] The negative effects of subsidies can be tackled by:

- Further subsidies, for example, agri-environment payments are made to increase environmentally beneficial practices, which are under-provided in part due to the effects of other agricultural subsidies;
- Changing the subsidies to retain their beneficial impacts while reducing the negative effects. Decoupling of CAP payments is an attempt at this approach;
- Reducing the subsidy, for example the process of modulation, which reduces direct CAP payments by a fixed proportion and diverts this money into rural development schemes, and the EU's 'financial discipline' which restricts the overall budget for agricultural support.

Reducing subsidies will decrease the use of resources in production, and therefore reduce associated negative externalities, leading to resource reallocation and less environmental disruption.[5] However, the total value of agricultural land to society is made up of a mix of different market and public values. As described above, agriculture results in a mixture of positive and negative externalities. Therefore, it is not straightforward to determine what level of agricultural subsidy (if any) is optimal for society.

The social value of the goods provided by agriculture can be analysed as a balance of the net value of production (which farmers might be expected to optimise in the given market conditions) and the supply of public goods such as rural employment, landscape preservation and diversity of wildlife. Society wishes to optimise the combination of these factors and (as each of these factors are positively or negatively correlated to agricultural production in different ways) the weighting between them determines the optimal level of subsidy.[4]

3 CAP, Decoupling and the Environment

The Biggest Subsidy in Europe

In the last decade, world agricultural subsidies have been estimated at over $300 billion,[2] or 1.3% of global GDP.[7] They are part of worldwide agricultural policies that generally aim to support incomes and stabilise prices.[7] The OECD estimate that of the $107 billion spent on supporting farmers in the EU in 2002, $61 billion of this came from consumers in the form of higher prices caused by tariff protection and export subsidies and $46 billion from tax transfers.[9]

The future Common Agricultural Policy (CAP) budget in the EU is planned to be €54 billion in 2006, taking 36% of the total EU budget. Within the CAP, rural development spending (which includes environmental spending) is planned to increase to €10 billion, taking around 20% of the CAP budget from 2006.[10]

Changing Objectives

In the last century, the CAP aimed to support more farming production than the market level, to ensure food security in the aftermath of World War II (as enshrined in the Treaty of Rome) and for reasons of achieving greater employment and economic development in rural areas than the market would deliver. If the level of farm production was closely related to rural employment and development, then subsidies for farm production might efficiently raise these to socially optimal levels.

Technological developments have also meant farm production has increased in tandem with large reductions in agricultural jobs; employment in farming has declined by over a third since 1970, and is still falling at 4% per year in England.[11] Farming now supports just 1.5% of gross employment in England,[6] and 6.5% of the rural workforce.[11]

The Policy Commission on the Future of Farming and Food[12] in England supported the idea of decoupling. They recommended 'the guiding principle (to CAP reforms) must be that public money should be used to pay for public goods...direct payments should be...decoupled from production and be subject to base environmental conditions'.

In future decoupling will mean payments will be linked to the respect of environmental, food safety, animal and plant health, and animal welfare standards, as well as the requirement to keep all land in good agricultural and environmental condition ('cross compliance'). However, member states have a large degree of flexibility in how they define and implement the reforms. Single farm payments will vary from partially coupled payments in France, to totally decoupled payments based on the historic subsidies a farmer received in the past in Scotland, Ireland and Wales, to the introduction of a flat rate area payment system in England and Germany. There is also an option for coupled elements to be maintained to avoid abandonment of production.

Environmental Impacts of the CAP

Environmental change is an inevitable by-product of agricultural activity.[7] The negative environmental and social effects of EU agricultural subsidies, and dumping excess EU farm produce on world markets, have been extensively analysed. They include the destruction of livelihoods in developing countries[13] and negative impacts on biodiversity.[14] Two of the most important environmental effects of agriculture on European wildlife are diffuse pollution and declines in farmland birds and other wildlife.

Agricultural and environmental policies in England are inextricably linked. Farming occupies 70% of England's land area,[6] and agriculture has significant impacts on the environment. Numerous studies have looked at the positive and negative environmental externalities of agriculture. Most of these impacts involve public goods, such as the existence value of nature and are external to markets, making them potentially suitable areas for government intervention.

In England and Wales, diffuse pollution of water by pesticides and nutrients from agriculture is recognised as a serious problem by the Environment Agency. Some £241 million of water company costs for remediation of raw water quality are attributable to external sources, such as agriculture.[15, 16] English Nature identify diffuse pollution from agricultural as the cause of unfavourable conditions on 13,000 hectares of SSSIs in England.[17] There have been substantial declines in farmland wildlife in the UK since the 1970s. In the 1990s, the population of farmland birds was adopted as an indicator of sustainable development by the UK Government.

These externalities relate to the intensity of agricultural production, rather than agriculture *per se*. Therefore, agricultural subsidies that encourage intensity of production are likely to exacerbate these environmental effects.

The CAP reforms agreed in 2003 involved only small changes in the overall level of agricultural subsidy. A potentially more significant change is in the structure of the subsidy payments: through decoupling, the subsidies will shift from farm output to farming *per se*. This will reduce the incentive for agricultural production, so should reduce both the intensity of agriculture and the associated externalities. However, decoupled subsidy payments will remain as an incentive for more land to remain in agricultural use than in a subsidy-free market.

Environmental Policy Responses

A typical environmental policy response in the UK, to counter the impacts of intensive production, is to extensify production. This can be done by regulations (such as the capping of farm inputs permitted in 'nitrate-vulnerable zones') or by incentive payments. Incentives are usually calculated on an 'income forgone' basis. This seeks to compensate the farmer for reducing the intensity of production, such as the maximum output per hectare farmed (*e.g.* restrictions on livestock units per hectare in Environmentally Sensitive Area schemes) or the proportion of the farm put to productive use (*e.g.* 'buffer strips' of land not used for production alongside sensitive features such as hedgerows or water courses).

Decoupling breaks the link between subsidy and production, and so is predicted to remove an incentive for farmers to maximise production.[18] It should therefore reduce the environmental damage from intensive farming, for example by reducing pesticide use and pesticide levels in ground and surface water.[19]

4 Case Studies of Changes to More Extensive Farming

The changes in the structure of CAP subsidy payments, through decoupling, could have a significant effect on the positive and negative externalities of agriculture. Decoupling the link between the subsidy payment and the intensity of agricultural production will reduce the incentive to produce agricultural outputs. This has important consequences for environmental policies that seek to address the negative environmental externalities associated with the intensity of agricultural production. A key parameter for these policies is what the size of change in the incentives will be, both in absolute terms and as a proportion of overall farm profitability.

To estimate the size of the reduction in the output incentive, we have constructed two simple case studies to examine the effects of decoupling on the farm business.[20] The relative costs of system changes to the farm business before and after decoupling are examined. A system change to extensify production (a typical environmental policy response discussed above) is considered.

Case Study 1

This first example looks at a hypothetical 200 hectare English arable farm. For simplicity we have assumed the same land area was farmed between 2000 and 2005 with similar cropping patterns in those years (see Table 1).

The cost of system change is estimated by calculating partial budgets for uncultivated land pre and post decoupling. In this example, production is extensified by leaving uncultivated land. The cost of this will be reduced after decoupling as the farmer is no longer required to produce a crop to qualify for subsidy payments. The impact of decoupling and thus the relative reduction in the cost of extensifying, arise immediately in 2005 irrespective of the 'phased' system of payments towards an area payment, which will be used in England.

The cost to the farm of leaving land uncultivated is calculated at £596 per hectare in a 'coupled' scenario such has existed until 2005[b] (see Table 2). In a decoupled system,

Table 1 *Assumed Crop Areas*

Crop	*Hectares*
Winter wheat (WW)	120
Oil seed rape	60
Set aside	20
Total	**200**

Table 2 *Farm loss for taking one hectare out of winter wheat production in 2004–pre decoupling*

Revenue Loss	*£/hectare*	*Revenue Gain*	*£/hectare*
Income foregone from having land taken out of wheat production*	861		0
Total	**861**	**Total**	**0**
Extra Costs	*£/hectare*	*Costs Saved*	*£/hectare*
Annual cutting to maintain agricultural condition**	15	All operations associated with production of an arable crop	250
Total	**15**	**Total**	**250**
Sub Total	**846**	**Sub Total**	**250**
Profit on change	0	Loss on change	596
Total	**846**	**Total**	**846**

* Based on gross margin of 8.2t/hectare at £75/t and with an arable area payment of £246.13/hectare
** One optional method of complying with good agricultural and environmental condition.

[b] This gross margin included is for winter wheat. When analysed for all crops on the farm the sensitivity was less than 10% and thus not included.

the income foregone from having land taken out of wheat production falls, because (provided cross-compliance conditions are met) the arable area payment of around £246 per hectare is no longer lost. As shown in Table 3, other things being equal, the predicted cost of leaving land uncultivated post decoupling is calculated at £350 per hectare. In a decoupled system, the loss on change therefore reduces by £246, or 41%.

The reduction in the cost of taking land out of production, post-decoupling, is because the farm does not incur the loss of the area payment in the decoupled system. The farm has been receiving arable area payments in the past and from 2005 this will be transferred into a partial historic decoupled single payment until 2012, when it will be a flat rate area payment. Between 2005 and 2012, CAP support payments are expected to fall steadily. Table 4 shows the total estimated support payments to the farm. It is based on estimated rates of modulation (10%), reduction for financial discipline (up to 10%) and the national reserve (3%). It shows expected total and per hectare payments for this 200 hectare arable farm.

Support payments to this farm are currently £246 per hectare. Following implementation of CAP reforms they are around £221 per hectare in 2005 and steadily reduce to around £162 per hectare in 2012. The change to an area based system may reduce farm income, dependent on the relationship between historical receipts and future area payments. There is also an expected overall reduction in support for two reasons. Firstly, the EU agriculture budget is put under further pressure through a combination of the 'financial discipline' limit and the EU's expansion continues to put this under more pressure. Secondly, modulation will also reduce the farmer's single payment and recycle it into rural development funds. Increases in support will be possible for every farmer in England, who will have the ability to obtain agri-environment subsidy payments.

Table 3 *Farm loss for taking one hectare out of winter wheat production in post decoupling (2005 and after)*

Revenue Loss	*£/hectare*	*Revenue Gain*	*£/hectare*
Income foregone from having land taken out of arable production	615		0
Total	**615**	**Total**	**0**
Extra Costs	*£/hectare*	*Costs Saved*	*£/hectare*
Annual topping*	15	All operations associated with production of an arable crop	250
Total	**15**	**Total**	**250**
Sub Total	**600**	**Sub Total**	**250**
Profit on change	0	Loss on change	350
Total	**600**	**Total**	**600**

* One optional method of complying with good agricultural and environmental condition.

Table 4 *Projected CAP Pillar 1 support payments to a typical English Arable Farm 2005–2012*

	2005	*2012*
Payment Levels		
Historic Rate	90%	0%
Flat Rate	10%	100%
Annual Payment		
Historic Rate	£ 43,634	£0
Flat Rate	£ 4,132	£ 41,316
Total Deductions	£ 3,557	£ 8,937
Total Payment Level	£ 44,208	£ 32,379
Payment per hectare	£221	£ 162

Calculated using ADAS Management Consultancy 'Single Payment' calculator

It is not possible to say what impact this will have on net farm income or farm profitability *per se*, as the CAP reforms will enable the farmer to change cropping or the farming system without altering their subsidy payment. The change to decoupled payments may also influence commodity prices through altering supply and leaving the producer free to react to market demands. Oil seed rape (OSR) is also likely to become relatively less attractive in economic terms compared to winter wheat post decoupling, because whilst the area payment has been the same (at around £246 per hectare) it has made up a greater proportion of the OSR gross margin than for winter wheat. The agronomic benefits of having OSR as a break crop will remain, however, and thus ensure the crop remains an important element of the farm cropping plan.

As the subsidy payment per area falls, the cost of taking land out of production remains the same, but the overall income of the farm business falls. Therefore the relative impact of the loss on the overall farm business may increase. This will need to be borne in mind when designing policy mechanisms post-decoupling.

Case Study 2

Our second example looks at a hypothetical 150 hectare English upland livestock farm. Again for simplicity we have assumed the same land area was farmed between 2000 and 2005 with similar livestock numbers kept in those years. In this example, a system change to extensify production through reducing livestock numbers is modelled.

The farm has 150 hectares of permanent pasture within a severely disadvantaged area of England. Some of the farm is rough grazing and the equivalent productive land area is calculated within the total of 150 hectares. Assumed livestock numbers are shown in Table 5.

Historically the farmer has received subsidies through a range of schemes such as the suckler cow premium scheme and the ewe premium, based on the number of livestock kept. The gross margins in Tables 6 and 7 show that after decoupling, when the farmer no longer has to keep livestock to claim the payment, it becomes less costly to reduce livestock numbers.

Table 5 *Assumed Livestock Numbers*

Stock	*Number*
Suckler cows	35
Steers	23
Breeding ewes	611
Other sheep 6 months +	257

Table 6 *Gross margin for one suckler cow pre and post decoupling*

	2004 £/cow	*2005 £/cow*
Calf sales	283	283
Suckler cow premium*	185	0
Less depreciation and calf purchases	–66	–66
Output	**402**	**217**
Concentrate	25	25
Puchased feed	9	9
Forage costs	67	67
Other variable costs	45	45
Variable costs	**146**	**146**
Gross Margin	**256**	**71**

* Includes suckler cow premium and extensification payments. Does not include the Hill Farm Allowance that the farmer will continue to receive in the future on an area basis as a supplement to the decoupled payment.

Table 7 *Gross margin for one upland sheep, pre and post decoupling*

	2004 £/sheep	*2005 £/sheep*
Lamb sales	45	45
Ewe premium	18.28	0
Wool	1.5	1.5
Less depreciation and calf purchases	15	15
Output	**49.78**	**31.5**
Concentrates	6	6
Forage costs	8.5	8.5
Other variable costs	9	9
Variable costs	**23.5**	**23.5**
Gross Margin	**26.28**	**8**

The gross margins show that in 2004 if the farmer made the decision to reduce stocking numbers this would cost in the region of £256 per cow. It is unlikely many additional savings such as fixed costs would occur in this scenario although labour costs may be slightly reduced.

The incentive in the past to reduce stocking has been through schemes such as Countryside Stewardship, which aimed to pay the farmer to make up for this loss of income. Post decoupling, if a decision is made to reduce stocking, in this example the cost will be reduced by £185, to £71 per cow. As Table 7 shows, similar conclusions apply to sheep production.

Before decoupling, a reduction of sheep numbers cost in the region of £26 per ewe. Post decoupling it is estimated this will reduce by £18, to a cost of around £8 per ewe. Therefore, the amount of compensation (on an income forgone basis) for the farmer to reduce stocking of sheep or cows is reduced by around 70%.

Table 8 shows that the impact of CAP reforms across the whole farm is a reduction in support payments, this being mainly due to the deductions for schemes such as modulation.

The two case study examples assume commodity prices will remain the same post decoupling. Although this is highly uncertain given the many different influences on prices in England, including the process of decoupling itself, which by removing the connection between subsidies and the quantities produced should increase the influence of markets on production decisions. No farm is ever 'average' and farming habitats and cultures have been learned from decades of previous practices. Furthermore, farmers do not always react as rational economic agents. Nevertheless, it is important to understand the change to the incentives, to help design environmental policies, and manage public spending.

Case Study Conclusions

The case studies show that the act of decoupling makes it cheaper to carry out environmental measures such as removing land from production or reducing livestock numbers. It also means that government incentives to carry out these practices should require lower payments, whilst the mechanism for compensation continues to be 'income foregone'.

If farmers act as rational economic agents every season they will analyse gross margins and make production decisions based on likely market returns within the

Table 8 *Projected CAP Pillar 1 suport payments to an illustrative English upland livestock farm 2005–2012*

	2005	*2012*
Payment Levels		
Historic Rate	90%	0%
Flat Rate	10%	100%
Annual Payment		
Historic Rate	£ 21,663	£0
Flat Rate	£ 1,697	£ 16,971
Total Deductions	£ 2,106	£ 4,020
Total Payment Level	£ 21,254	£ 12,952
Payment per hectare	£142	£ 86

Calculated using ADAS 'Single Payment' calculator.
The farm will also continue to get Hill Farming Allowance. As with the decoupled single farm payment, this is separate from per animal livestock support payments, so is not included within the calculations.

context of their whole businesses. This could lead to more cyclical production of agricultural produce as farmers choose to reduce or not carry out any productive activity when it does not offer them a satisfactory profit. In short, the decision-making process becomes far more like those of other businesses, within the parameters of environmental compliance and with the Single Farm Payment acting as financial security to the individual producer. However, this decision-making process is unlikely to occur in the short term and many producers will continue to 'couple' the single payment in their minds.

5 Implications for Farming Policy

Farm-Level Policy Impacts

In general, environmental assessment of the benefits of decoupling is extremely difficult, as the effects of markets, policies and environmental responses are difficult to predict. The two case studies estimated that decoupling would reduce the costs of extensification significantly:

- The cost of taking arable land out of production is reduced by £246, to £350 per hectare (a 41% smaller loss compared to the present coupled system);
- The cost of extensifying grazing is reduced by £185, to £71 per cow, and by £18, to £8 per ewe (a 70–75% reduction in the loss compared to the present coupled system).

These reductions in the costs to the farm business of extensifying production have significant impacts on environmental policy design. For example, if a scheme wanted the farmer in Case Study 2 to reduce his suckler cow numbers by 25%, this would have cost the farm £2,304 in the old coupled system. After decoupling, the cost of this change is £639.

In recent years, high costs to farmers have been one major factor, amongst others, for not introducing greater environmental policy measures (such as a pesticides tax or wider designation of nitrate vulnerable zones) in the UK. Following decoupling, the cost of adjusting to these policies, through more extensive production practices, should be significantly lower.

However, decoupling does not mean that subsidies paid to farmers will be completely stable. Our case studies show that between 2005 and 2012, the total subsidy payment per area falls, so the overall income of the farm business falls. The cost of taking land out of production remains the same over this period, so the relative impact of the loss on the overall farm business increases. This will need to be borne in mind when designing policy mechanisms in England until 2012.

In addition to environmental subsidies, decoupling may make other environmental policy instruments more attractive. The World Bank identifies that in order to use markets in environment policy, persistent market distortions need to be removed, for example, by reducing subsidies.[2] OECD analysis of the CAP reforms show that the key dividing line is between policies that primarily distort markets and policies that interfere less with market forces whilst also offering a better chance of achieving other important policy objectives in an effective way.[9] Therefore, decoupling may make the use of markets in environmental policy (for example, by promoting foods produced in an environmentally beneficial manner) more effective.

Impacts on Public Spending

Decoupling should significantly reduce the cost of environmental options for individual farmers. The same, therefore, might be expected in terms of overall costs to public spending of environmental options; the lower income forgone after decoupling means that the gross cost of an environmental payment should be lower per hectare.

These lower environmental payments are a result of compensating a much lower reduction in subsidy payments. Previously this reduction in subsidies would also have created a gross saving to public spending, making the net public costs of environmental payments for decreased production significantly less than their gross costs. For extensification payments after decoupling, just as the gross cost to the farmer falls, so the gross saving to public spending on direct payments is also reduced. Therefore, the change to decoupling does not offer as big a net saving to public budgets as might first appear.

However, decoupling does make payments more predictable. Single Farm Payments will not vary with the level of production. The large differential between the gross and net costs of environmental policy measures will be removed. This should make the costs of environmental policies more transparent and therefore easier to manage.

Potential Negative Environmental Consequences

There are also potentially negative environmental consequences of decoupling. For example, it is unclear what the effects of reducing subsidies linked to production will be on farmers in economically marginal but ecologically valuable land. In these circumstances, calculating environmental payments at the margin, through the income forgone system, may not be adequate. Payments that aim to maintain agricultural systems where they are of particularly high value to society may also need to consider the viability of farm businesses and effects of reduced CAP payments. Therefore, environmental subsidies may need to be calculated on a total or average cost basis, to support whole farm businesses.

6 Conclusions

In our two farm business case studies, designed to represent typical arable and upland livestock farms in England, the costs of changing to more extensive agricultural production are reduced by 40–70% following decoupling. The consequences of this are that the compensation paid to farmers to take these steps is significantly lower, and will have a more predictable impact on public expenditure.

For environmental policies in agriculture, previously rejected economic instruments (such as taxes) and alternative measures (such as market mechanisms), may now have lower costs for the farming industry as a result of decoupling. Therefore, policy choices made under the old coupled system may no longer be justified. In the longer term, the Government's farming strategy will need to be updated to reflect these changes in economic incentives.

Acknowledgement

The authors would like to thank Robin Smale, of Oxera Consulting, discussions with whom initiated the idea for this article, and Dr Sue Armstrong-Brown, of the RSPB, for useful input to its development. However, responsibility for content lies solely with the authors.

References

1. Cabinet Office, http://www.cabinet-office.gov.uk/regulation/role/index.asp, 2004.
2. D. Pearce and E. Barbier, *Blueprint for a Sustainable Economy*, Earthscan, London, 2000.
3. G. Bannock, R. Baxter and E. Davis, *Dictionary of Economics*, 5th Edn, Penguin, London, 1992.
4. G. Edwards-Jones, B. Davies, and S. Hussain, *Ecological Economics*, Blackwell, Oxford, 2000.
5. NIEIR, *National Institute of Economic and Industry Research: Subsidies to the use of natural resources*, Australian Department of the Environment, Sport and Territories, updated.
6. GHK & GFA-RACE, *Revealing the Value of the Natural Environment in England*, Report to DEFRA, 2004.
7. J. Lingard, Agricultural Subsidies and Environmental Change, in *Encyclopedia of Global Environmental Change*, John Wiley and Sons, Ltd., Chichester, 2002.
8. D. Guellec and B. van Pottlelsberghe de la Potterie, Does Government Stimulate R&D?, *OECD Economic Studies* **No 29**, 1997/II.
9. OECD, *Analysis of the 2003 CAP reforms*, OECD, Paris, 2004.
10. European Union, *Financial Framework for 2007–2013: Analytical Report.* Council of the European Union, Brussels, 2004.
11. Countryside Agency, *Rural Economies*, Countryside Agency Publications, Wetherby, 2003.
12. The Policy Commission on the Future of Farming and Food, *Farming and Food: A Sustainable Future*, DEFRA, London, 2002.
13. Oxfam, *Stop the Dumping*!, Oxfam, Oxford, 2002.
14. RSPB, Eat This: fresh ideas on the WTO Agreement on Agriculture, RSPB, Sandy, 2001.
15. ERM, Stone & Webster, Assessing current levels of cost-recovery and incentive pricing, DEFRA, London, 2004.
16. J.N. Pretty, C. Brett, D. Gee, R.E. Hine, C.F. Mason, J.I.L. Morison, H. Raven, M.D. Rayment and G. van der Bijl, An assessment of the total external costs of UK agriculture, *Agricultural Systems*, 2000, **65**, 113–136.

17. English Nature, Website, April 2004: http://www.english-nature.org.uk/special/sssi/ reportAction.cfm?Report=sdrt17&Category=N&Reference=0
18. DEFRA, Regulatory Impact Assessment on Options for the Implementation of the Reform of the Common Agricultural Policy, DEFRA, London, 2004.
19. GFA-RACE and IEEP, Impacts of CAP reform agreement on diffuse water pollution from agriculture, DEFRA, London, 2004.
20. Based on data in J. Nix, *Farm Management Pocketbook*, The Andersons Centre, Melton Mowbray, 2004.

Globalising Vulnerability: The Impacts of Unfair Trade on Developing Country Agriculture

JAMES SMITH

1 Introduction: Linking the Global to the Local

The self-immolation of Lee Kyung Hae at the World Trade Organisation (WTO) Ministerial Meeting held in Cancún, Mexico, in September 2003 threw the impacts of unfair trade on developing country small-scale farmers into sharp relief. Mr Hae, who was wearing a sign that stated 'the WTO kills farmers,' died at the security barricades that separated the protestors and activists from the conference centre that housed the WTO. An estimated 10,000 activists marched in Cancún, and an estimated 10,000 Mexican police were stationed there to ensure the march had nowhere to go. The barricades represented not only the political gulf that exists between those who are most severely impacted upon by the realities of international trade rules and those who are charged with regulating those rules more fairly, but also the complexity of mapping the many different routes through which seemingly disconnected circuits of trade impact on the small-scale farmer and agricultural practices in less developed countries. It can be difficult to analyse and assess the seemingly intangible interconnections that link farmers who often do not even enjoy the tangible connections of mains water, electricity and all-year-round transport links, to the dense, arcane discussions that shape international trade in Geneva, Cancún and elsewhere. Equally, the protestors who reach the inevitable barricades at these destinations find it difficult to believe that the people with the power to regulate the shape of international trade give any thought at all to the impacts of their decisions on farmers in less developed countries. Such seems to be the reality of the politics of globalisation.

Despite the evident disarticulations that exist between the realities of less developed country agriculture, the mechanics of international trade and the politics that frame the debate, an international movement has gained momentum to try to alter

Issues in Environmental Science and Technology, No. 21
Sustainability in Agriculture

the way trade is practised, to change the way farming is organised in the northern hemisphere for the good of the southern hemisphere. The global reorientation of agricultural practices and the ways agricultural commodities are traded that is required for a fairer balance to be struck is potentially massive; one billion US$ per day is currently spent on agricultural subsidies by the more developed countries, with enormous implications for competitiveness and international commodity prices.[1] These enormous sums of money support entire farming industries in the richer countries and keep powerful farming lobbies agreeable. There are many political and economic stumbling blocks to making international trade fairer for developing countries, but the benefits are also potentially staggering. Several international non-governmental organisations (NGOs), including Action Aid, Oxfam and Christian Aid, have attempted to aggregate the potential benefits of fairer trade, and these potential benefits certainly outweigh the current value of development aid and other development assistance.[1] As protestors and activists point out, the desire for fairer international trade practices is not only based on moral grounds, but based on practical and financial realities too. Political inertia remains, however.

The political debates that surround the issue of unfair international trade and its impact on developing country agriculture are vocal, charged and well documented. The real, concrete impacts of unfair trade on agriculture, rural livelihoods and food security in less developed countries are much harder to ascertain.[2]

Anthropologists, geographers and development studies specialists have long tried to link the realities and rural livelihoods of small-scale farmers in developing countries to the array of contexts and constraints that shape the livelihoods they are able to construct for themselves. It is useful to sketch out some of these central ideas before we begin to explore the impacts of unfair trade on developing country agriculture. In the 1960s, anthropological studies began to highlight the adaptive, complex livelihoods of indigenous communities. Indigenous communities, who had previously been perceived as environmental illiterates, were understood to have a thorough understanding of the environment in which they lived, and engaged with it through an array of complex productive mechanisms.[3] A fundamental weakness of this perspective was the narrow focus that it employed. The community and household level research-focus anthropology employed did not contextualise communities and households within broader political and economic frameworks. However, interest in how communities and their resource use were being integrated and transformed through the influence of a global economic system was growing and a broader perspective proved necessary to analyse this. In response a new, somewhat Marxist-inspired form of analysis emerged: political ecology.

The discipline of political ecology is central to any meaningful analysis of human–environment interactions. Political ecology is in part a rejection of the Darwinian and Malthusian ideas that dominated environmentalism in the 1960s. Political ecology emerged as a response to the need to integrate the study of land use practice within a broader political economic framework. Work by Blaikie forms the theoretical core of political ecology.[4] Political ecology seeks to place the concerns of ecology within a political economic framework. In accordance with this, soil erosion in less developed countries is not necessarily a result of overpopulation, ignorance and bad management practice but rather one of political economic constraint. Central to this analysis is the 'land manager' who must be considered within a historical, political and economic context. Political ecology is in essence concerned with relations of production–access to and control of environmental, or productive, resources. The argument runs that a lack of access to environmental resources, marginalisation, is self-reinforcing. Marginalisation is both a cause and a consequence of a lack of access to environmental resources. Social relations place excessive production pressures on environmental resources, which may be transmitted back to the land manager in the form of environmental degradation. Political ecology turns on the assertion of two key statements: first, marginality is a result of political and economic constraints; and second, marginality, rather than population numbers, puts pressure on the environment.

Political ecology, then, allows for a perception of the resource-poor, small-scale farmer constrained by a variety of political, economic and cultural factors that may operate at a range of levels, from the local to the global. This broader perspective of what constrains the decisions a small-scale farmer may make, and consequently what shapes his or her agricultural practices and livelihoods, hints at the possibility of an array of articulations between the small-scale farmer and the increasingly global economy in which he or she operates. Where political ecology often falls down, however, is in analysing these relationships in anything other than an intuitive manner. For example, if a small-scale farmer is unable to access international commodity markets in a fair manner, intuitively it suggests that farmer is losing out on potential income. There are, however, a range of other factors that need to be considered; for example, would that farmer have access to international markets even if he could trade fairly on them or would issues like a lack of information about how the market would work or an inability to transport commodities mean he was unable to enter that market anyway? Would the farmer choose to cultivate a crop that he could trade internationally if the market was open to him or would he rather choose to grow a mix of subsistence crops and cash crops? Would the farmer face new constraints to his agriculture that would prohibit his ability to exploit an international market, for example a lack of family labour or increased competition from his neighbours? There is another layer of relationship between small-scale farmers, their ability to farm and the markets they operate within: individuality and choice. The choices a small-scale farmer may make may not appear rational in purely economic terms; individual preferences and cultural norms mix to shape the way a farmer may see his or her land, and the possibilities it contains. Making trade fairer may for one farmer be the opening up of a world of possibility and profit, whilst a neighbouring farmer may not see or be interested in this possibility.

The nub of this is that whilst we can intuitively gauge what impacts macro, global rules and regulations may have on the micro level of the community and small-scale farmer, we can rarely assess with any certainty the real or potential impacts on small-scale farmers in a way that allows us to gauge the scale of an impact and the trend of impacts in the future. It may be clear in particular cases that unfair trade has significantly impacted upon an individual, for example a farmer in Kenya who has been forced out of coffee production, but that does not make it easy to scale that back up to assess the scope of an impact. It is clear that unfair trade does significantly constrain the range of opportunities small-scale farmers may have to construct their living, but agriculture in developing countries is a complex business, constantly evolving, spread across many physical environments, focused on many different commodities, shaped by a variety of cultural practices and norms, and developed within particular sets of individual and national histories. The trick is to draw on the broader-scale analyses of unfair trade, and make use of pertinent case study material to provide a judicious sketch of how unfair trade impacts upon agriculture in developing countries. That is the aim of this article. The first part of the article will present a very simple overview of the relationship between international trade and less developed country agriculture at the broader scale, highlighting the role of subsidies, tariffs, trans-national companies and trade rules that shape the uneven topography of international trade. The second part of the article will focus on the ways in which rural livelihoods and less developed country agriculture can be impacted upon by international trade, with an emphasis on the complexity and variety of agricultural systems and rural life in less developed countries. The third and final part of the article will focus on the tangible impacts of unfair trade in developing countries, using a case study of sugar production in southern Africa, and in particular South Africa, Malawi, and Mozambique, and their relationships with the European Union. Sugar is chosen as a commodity, and southern Africa is chosen as a discrete region, as these case studies present a whole range of impacts, across a whole range of different types of farming activity, and highlight issues that strike to the core of unfair trade and less developed agriculture. The article concludes by attempting to synthesise the practice of international trade with the impacts on agriculture in an aim to illustrate how unfair trade constrains agriculture, and by extension rural development, in less developed countries.

2 An Overview of International Trade and Less Developed Country Agriculture

Within the term unfair trade lie two obvious questions. First, what do we mean by unfair trade, and second, what are its implications? This section will attempt to answer the former question and the following section the latter. Unfair trade is actually a compound of several different interlocking trade regulations, practices and shifts in the global agro-food industry. Subsidies for agricultural commodities produced in more developed economies, tariffs that prevent less developed country exporters from easily accessing the economies of more developed countries, and the increasing dominance of trans-national companies over the production, processing, and marketing of agricultural commodities dovetail to favour more developed country farmers heavily in the international marketplace over less developed country farmers. It is important to point out that the impacts of these developments are not uniform across less developed countries; some of these countries enjoy more preferential trade agreements than others, and often such countries have historical or colonial links to particular more developed nations. Indeed, one of the weaknesses of the WTO is that it is undermined by an array of privately negotiated bilateral trade agreements over which it has little control or authority. Further, unfair trade may be practiced between less developed countries; very high tariffs exist between many African countries for example, and South Africa has a history of dominating trade over certain commodities in southern Africa. The realities and impacts of international trade cannot be simply mapped as a region of more developed countries who dominate and prosper from trade and a bloc of less developed countries who do not; the rules, regulations and agreements that shape international trade form an extremely complex regulatory matrix that shapes equally complex subsets of benefits and burdens. Attention now turns to examining some of the key unfair trade practices.

Trade and Globalisation in the 21st Century: the Political Context

The idea of free trade is wrapped up in ideas of political and economic integration and globalisation. The argument runs that the removal of barriers to international trade will create a system whereby countries can more effectively utilise their comparative advantages in order to capture profits and capital. This perspective, which allows for the fact that aggregate gains will be greater for the already wealthy, holds that sufficient gains can still be made for poorer people in less developed countries.[5] This so-called 'trickle down' approach has been the cornerstone of thinking around globalisation from a developmental perspective for several decades. It is not uncontroversial, however. The idea of globalisation as a benign inevitability that will eventually lift people out of poverty has been challenged on several fronts. Studies have shown that the poorest and most vulnerable sectors of society in less developed countries become relatively significantly poorer upon integration.[6] International inequality has increased significantly since the early 1980s, and intra-national inequality has also increased within the majority of countries, and in particular the poorest countries. Many people in such societies cannot benefit from the potentials of globalisation as they are not linked in any meaningful way into markets, or do not produce commodities that could bear them a profit in markets. Furthermore, elites tend to control access to a majority of productive resources; thus it follows that, even in conditions of free and fair trade, aggregate gains amongst the resource-poor will be small. An array of factors constrain the ways in which people can interact within global markets even if trade were fair.

Globalisation is often perceived and packaged as an historical inevitability, a process that will occur outside of human intervention:

> [Globalisation] constructs the present as a moment, which is part of a fundamental historical transformation. Globalisation has become the grand narrative which justifies the end of all other master narratives of social change.[7]

Globalisation, in which free trade is effectively implicit, is often seen as an apolitical process. Thus, whether people can or cannot take advantage of that process is not the fault of politics but is a consequence of poor governance, limited access to resources, or an unwillingness to integrate or engage. This perspective, coupled with the perspective that the trickle-down approach will eventually provide material gains for all, places the fault for an inability to benefit from processes of globalisation and international trade firmly at the feet of those who do have the resources to benefit.

A more critical perspective of globalisation would argue that it is in fact a highly political process. Far from being an historical inevitability, globalisation has been politically poked and prodded into existence; far from having a life with an evolutionary pathway of its own, globalisation is the product of political alliances and agreements, rules and regulations, and networking and possibly coercion.[8] The politics of globalisation is not so much about integration for all but is more about selective inclusion and exclusion. First, some people are more able to engage effectively in a globalising world than others. Second, particular countries and people wield differing amounts of power when negotiating their way through globalisation and integration. Third, it is important to recognise that a rapidly globalising world and the associated paraphernalia of international trade has been created by us. Beliefs, alliances, values, norms, interests, and ideologies are deeply embedded within the global networks that we create. Globalisation and international trade are highly politicised, benefitting people with the power to shape them. It follows that debates surrounding international trade and whether it is fair or not are deeply political. Witness Cancún. For now, we briefly turn our attention to some of the ways in which international trade is practiced and why it may be seen as unfair.

Subsidies and Tariffs

Subsidies and tariffs can be regarded as the architecture of unfair international trade. Subsidisation of industrial country agriculture works in two main ways that discriminate against less developed country agriculture. First, the subsidisation of primary commodities such as sugar makes less developed country exports into subsidised markets uncompetitive. Second, widespread subsidisation serves to artificially depress the international market price of commodities such as sugar, which eats into the profit margins of unsubsidised farmers or may even push the international market price below the cheapest possible cost of production. For example, World Bank data shows that the cost of production of heavily subsidised European Union (EU) sugar exporters is, without subsidy, over four times the world market price. Countries like Swaziland and Mozambique, despite being two of the cheapest producers in the world, are able to produce their sugar only marginally below this artificially lowered price. The imposition of subsidies, however, means that these African exporters cannot compete. Meanwhile, the highly uncompetitive EU trading bloc accounts for 14% of all exports.[9] EU subsidisation of sugar does not only serve to give European sugar producers and exporters an unfair advantage, it plays a role in driving the volatility of the international sugar market. The market price of sugar has suffered a trend of decline over the past two decades, punctuated by short, sharp price rises and falls. Oxfam, who have conducted a large amount of research on the impacts of the EU sugar production regime, assert that despite the fact that the *value* of sugar trade has remained fairly constant since the mid 1990s, the *volume* of sugar produced has increased by 75%.[10] Production outstripping supply to such an extent has led to declining commodity prices and sugar exporters have been forced to increase the size of their exports in order to maintain their profits in real terms. Of course, increasing imports leads to a further depression of international prices. There are several reasons for this oversupply, including Brazil's emergence into the market in the late 1990s. The EU employs a system whereby it guarantees prices to EU producers; the quotas currently in place are far in excess of what needs to be produced to guarantee self-sufficiency, so a structural surplus is created that is effectively dumped onto the international market. The subsidisation of EU sugar producers, therefore, continues to play an important role in dampening international market prices.

To continue with sugar as a commodity, the EU's subsidisation of sugar is not the only impact on sugar as a commodity. Less developed country exporters find themselves caught in a double movement when trying to export to the EU. First, there is the heavily subsidised EU sugar to compete with on international markets,[11] and second, there is the array of tariffs to overcome before sugar can be exported into the EU trading zone. In addition to a fixed tariff, a special safeguard is employed. The fixed duty is set at €419 per tonne, and if international sugar prices fall below the level of €531 per tonne an additional duty is applied. As Oxfam point out, this safeguard has been in place since 1995. EU sugar producers, then, are supported and protected by a matrix of reinforcing policies that guarantee prices, protect the internal market, and subsidise exports. The figures are enormous; each year European consumers and taxpayers foot a bill of US$1.57 billion for the full gamut of subsidies and tariffs.[12] Large sugar producers in the UK receive subsidies of almost US$100,000 annually, that is almost 400 times the average Gross Domestic Product *per capita* in Mozambique. British Sugar, one of the largest firms in the sugar processing industry, receives around US$100million annually from the Common Agricultural Policy (CAP), accounting for over half its profits.[13] The EU often justifies CAP and the sugar regime as an exercise in rural development, maintaining food security, and as an environmental benefit. This, of course, was the reasoning behind the original conception of CAP, but CAP no longer exists in the post-World War Two context. CAP is a production-oriented network of subsidies–the agricultural sector is subsidised to produce whether a market exists or not, although movements within Europe, such as its recent expansion, will alter the course CAP takes in the future. The inevitable consequences of such a policy orientation are an increase in the gap between richer and poorer farmers, the encouragement of inefficient resource use, and the support of environmental degradation. However different they may be in terms of production, organisation and rationale, there are distinct linkages between farms in more developed and less developed countries. The impacts in southern Africa are created by a situation where local producers see their exports barred through EU tariffs and face unfair EU competition in third markets.[14] The reality of sugar production is quite clear. Developing countries can produce sugar far more cheaply than European producers. Sugar grows particularly well in southern Africa and is an important crop for small-scale farmers across the region. It is a commodity in which southern Africa has a significant comparative advantage over European producers. Europe should be importing its sugar. Unfortunately, an array of subsidies means that the EU, one of the world's highest cost producers of sugar, is the world's largest exporter of sugar, accounting for 40% of world exports last year.[15]

In summary, developing countries are impacted upon by Europe's sugar policies in four key ways:

- *Restricting market access.* High tariffs and import quotas prevent some of the world's poorest countries from gaining access to EU markets, with attendant losses for rural incomes, employment and foreign exchange earning. As a result, Oxfam estimates Mozambique has lost the chance to earn an estimated

US$106million by 2004. That's almost three-quarters of the EU's annual development assistance to Mozambique.[16]
- *Undercutting export opportunities.* The dumping of European sugar overseas pushes other exporters out of third world markets. For example, opportunities for southern African producers are lost in Nigeria and Algeria due to European dumping in these countries.
- *Undermining value-added processing.* A handful of African, Caribbean and Pacific (ACP) developing countries receive some quota access to export their cane sugar to the EU but it is only raw sugar. Processing takes place in Europe. This inhibits the development of the sugar processing industry in developing countries.
- *Depressing and destabilising world prices.* Europe subsidises its sugar exports to bridge the gap between its own high price and low world prices. Even if world prices fall unsustainably low Europe subsidises the difference. The EU depresses world prices, often to levels below the costs of production of even the lowest cost producers such as Malawi, Mozambique and Zambia. This shrinks the foreign exchange earning potential of these countries. A lack of foreign exchange is one of the primary reasons why these countries have been unable to import emergency foodstuffs in the current southern African famine.

Sugar is just one example of a commodity that is traded in such a way as to accrue benefits for farmers in more developed countries to the detriment of farmers in less developed countries. It is, however, indicative of the range and extent of subsidies and tariffs that are employed to support relatively few people's livelihoods in the north (albeit people with a powerful lobby) to the detriment of many millions of people in the south. There are other shifts besides the algebra of subsidy and tariff that shape unfair agricultural trade in both the north and the south. We turn our attention to the role of agro-food restructuring and, in particular, trans-national companies.

Agro-food Restructuring

Trans-national companies are one of the key factors and one of the key articulations with respect to globalisation and international trade; they are a key component in a whole shift in the ways in which food is produced, manufactured, marketed and consumed. The ways in which we consume food are changing; we demand organic food, exotic fruits, or increasingly fruits and vegetables out of season. The removal of seasonality from the supermarket means that supermarket purchasers must increasingly look globally in their search for produce.

Meanwhile, disparities between those who are hungry and those who are not, and what the rich and poor eat, *within* and *between* countries both north and south, are increasing. Goodman and Watts talk of a global 'cool chain' that shuttles increasingly sophisticated agricultural produce around the world.[17] This global cool chain is both a cause and a consequence of massive restructuring of the food economy at the levels of production and consumption. Trans-national food companies and retailers are increasingly driving the transformation of the agro-food economy. In Goodman and Watts, Harriet Friedmann asserts that these trans-national companies are the major players in attempting to regulate agro-food systems; organising stable conditions of both production and consumption, which allow for the planning of investment, the sourcing of agricultural materials and marketing on a global scale.[18]

The machinations of trans-national companies such as Nestlé or British American Tobacco are increasingly felt in less developed countries, mainly through sub-contraction of agricultural activities. Sub-contracted activities tend to employ the most vulnerable workers such as women and children and are generally not encumbered by regulations regarding working conditions, job security and earnings. Where international trade in agricultural products does produce benefits for workers in less developed countries these benefits may not be as great, secure or sustainable as they could or should be. This, too, is true of the sugar industry in southern Africa where two or three processors dominate production through a system of contracts in South Africa and Malawi. In Mozambique, the re-emergence of the private sector has led to worries over working conditions as the private sector very reluctantly agreed to take over responsibilities for housing, services and healthcare from the previously state-run sugar companies. Issues over working conditions remain, however.

Trans-national companies also play a major role in shaping the topography of international trade. They connect production, processing, distribution and marketing in new ways. Their global reach means to a large extent they are self-regulating and as potential importers of foreign investment and creators of jobs developing countries are often happy to relax regulatory systems regarding employment, welfare and environmental protection in return for a company moving in and setting up.[18] Trans-national companies are able to negotiate positions for themselves that allow them to take advantage of cheaper labour costs and production opportunities through their financial clout and influence. Meanwhile, sub-contracting continues to place the risk of failed crops onto the farmer. The profits are made much further up the supply chain where value is added; less developed countries tend to be used as sources only of primary production and hence (partly as a result of import tariffs that encourage unprocessed commodities to the detriment of processed commodities), as we have seen with sugar, potential profits continue to be based on the vagaries of uncertain and tightly-margined international markets. The globalisation of agro-food production systems slots nicely alongside the tenets of free trade, but the realities of contracts, unreliable labour and unregulated labour conditions impact severely upon the way rural people farm and construct their livelihoods in the developing world.

Less Developed Country Agriculture and the Global Trading System

So far we have briefly looked at some of the mechanisms that are used to create an uneven playing field for international trade in agricultural products. Export subsidies, import tariffs and price guarantees are implemented, primarily by the EU and the United States, to secure the viability of more developed country industrial agriculture. Trans-national companies, too, play an important role in shaping the evolution of agro-food production; risk continues to be placed in less developed countries, meanwhile profits gained through processing and marketing exercises tend to accrue in the more developed countries where the trans-national companies are based. It is quite wrong-headed to think of one process of globalisation taking place. Globalisation in the more developed economies is controlled, regulated, and managed and where the agricultural sector is at risk from competition, more often than not, state support is put in place. Meanwhile, the less developed economies are put under considerable external pressure to submit to more free trade. Conditionalities for loans and putting loan repayments into abeyance unerringly call for less developed countries to relax or remove their own trade restrictions. Since these countries cannot afford to subsidise their own agricultural activities, their farmers are left further exposed to the globalisation of the agro-food sector.[19] These decisions are power-laden and are made internationally and until the beginning of 2004 the WTO tended to overlook agricultural trade due to an earlier gentleman's agreement. And that is precisely why protestors such as Lee Kyung Hae feel such impotence and frustration at the way in which international trade is regulated, particularly in light of the enormous impacts that unfair trade can have on developing country agriculture. That is the focus of the next section.

3 Impacts of Unfair Trade on Developing Country Agriculture

So far we have examined some of the instruments that are used to create an uneven playing field for international trade. We also looked at some of the conceptual difficulties in performing macro–micro analyses of the impacts of unfair trade on less developed country agriculture. A detailed, multi-country case study is employed to study the impacts of unfair trade policies on specific commodities and assess the implications for developing country agriculture. Before we turn our attention to that, however, it is important to underline the complexity and context-specificity of less developed country agriculture and rural livelihoods. Rural life in less developed countries is deeply complex. Rural people draw on a wide array of resources in order to construct livelihoods.[20] Natural resources, labour, social relations, tangible (goods, capital) and non-tangible (kinship, favours) assets are all utilised in order to survive. By knitting themselves to a complex web of resources, however, people are making choices that may limit their ability to survive and flourish in the future. Rural livelihoods may be constrained and supported by externalities at a range of levels. At the global level, ideologies and policies may limit access to resource bases, at the household level social hierarchies within households may control and contest the availability of labour.[20]

This diversity and reflexivity of rural livelihoods is both a strength and a weakness.[21,22] Intuitively, diversifying livelihoods through a broad portfolio of productive activities allows livelihoods to survive a shock or transformation in a particular area of the resource base.[23] During periods of drought, for example, people may diversify and seek wage labour. Diversity, however, can also become a weakness. If livelihoods are dependent on a wide range of activities, they become vulnerable to a broader spectrum of changes. Rural livelihoods become increasingly context-bound and locked within broader transformations and processes–political, economic, social and environmental.[24] With this in mind, Richards[25] conceptualises rural livelihoods as a 'performance . . . not a design, but a result', constantly being reconfigured and renegotiated through space and time. The ways in which this 'performance' is shaped are complex and nuanced, determined not only by what access to resources people have, but also by a desire to maintain and improve their way of life.[26] The reflexive manner in which rural livelihoods are constructed is to a certain extent an expression of the broader contextual changes that frame rural life. Rural livelihoods cannot simply be analysed as discrete units, divorced from the forces that constrain and drive them, or analysed as if frozen in time. They are diverse and dynamic and 'performed' within the dynamic array of processes, rules and realities that mould, shape and constrain them. Rural livelihoods are the specific results of complex interactions with resources, opportunities and constraints at an array of levels. The seeming disconnection, yet inherent connection, of international trade and small-scale farmers in sub-Saharan Africa is a case in point. Yet, whilst acknowledging the complexity of individual livelihoods and the myriad contexts in which they are constructed, we need to search for a frame of reference in order to make our own conceptual connections between unfair trade and less developed country agriculture. With this in mind it is perhaps useful to think about three idealised agricultural/livelihood hybrids that encompass an enormous array of agricultural realities and potential impacts (see Table 1).

The first model is of a farming system that focuses on domestic, or subsistence, consumption. An agricultural system of this sort is unlikely to be affected much by unfair trading practices as the majority of agricultural production is consumed within the household or perhaps exchanged within the locale for other goods. This type of farming system is fairly closed in that it is not connected to circuits of international agricultural trade and therefore does not bear any of the benefits or burdens.[27] The second model of farming system is one focused on export. This encompasses a range of farm sizes and structures, from the relatively small 'emerging' farmers expected to develop in the larger developing country economies such as South Africa[28] through to larger agro-industries, perhaps more akin to factories, that produce sugar in Mozambique or coffee across Latin America, for example. This group of agricultural systems are most directly affected by unfair trade as their export opportunities

Table 1 *Three agricultural activity/rural livelihood models*

	Domestic consumption	*Export*	*Multiple livelihoods*
Attributes			
Activities	Farming for household needs	Some subsistence and production for local and/or other markets	Some farming/wage labour/remittances
Linkages to local market	Very few	Purchasing and selling	Purchasing
Linkages to other markets	None	Selling: via contract or middleman	Through employment
Potential impacts of unfair trade			
Constraint of opportunities	Very few	Reduced access to international markets	Very few
Sale of commodities	Very few	Reduced pricing due to dumping	Very few
Labour constraints	N/A	N/A	Poorly paid, seasonal, little regulation

may be constrained in terms of a lack of access to international markets and the decline in primary commodity prices.

Recent data collected in Kenya illustrates the extent of the problem as large decreases in coffee production due to low international prices, poor husbandry and contractual wrangling forced many farmers to search for new commodities to grow (coincidently this has provided opportunities for technologists to promote the use of plants generated using new technologies such as tissue culture).[29] Furthermore, agro-food restructuring may mean that many of these farms are caught in risk-laden and benefit-poor contracts within the increasingly global commodity chains described by Harriet Friedman.[18] The third class of 'farming system' may not incorporate much, if any, farming at all; a multiple livelihoods system is meant to represent the many rural households which rely on a wide array of productive activities to survive. Elements of agriculture may be incorporated (and many of these elements may be shrinking in importance)[30] but rural wage labour and urban remittances of cash and goods are also important. In rural Kenya, for example, a reliance on ten or more productive activities within a single household is not uncommon.[31] In this class of 'farming' system the impacts of unfair trade are likely to be felt when working on other people's land or in their factories. The reintegration of South African maize production, coupled with internal deregulation, led to the loss of 300,000 jobs in the late 1990s.[32] The case study of Mozambican sugar production presented in the next section also illustrates the realities of shifting working conditions.

It is important, too, to recognise the linkages between different types of agricultural production. Wage labour earned on commercial, export farms may play an important role in injecting capital into a rural area, and the knock-on benefits of this may be felt far beyond just the immediate or extended family. Similarly, the loss of wage labour may be mitigated against by owning land on which subsistence agriculture can be practiced or engaging in a range of other activities. In some senses, all households and certainly all communities are based on and reliant on a wide range of different activities. It is in untangling these connections that the true impacts of global processes such as unfair international trade can be discerned. The intent of this section was simply to illustrate the complexity, reflexivity and diversity of rural livelihoods in the developing world. It is within this context that we must attempt to analyse the impacts of unfair trade. This will be done through the presentation of a detailed case study, an analysis of the impacts of unfair trade on sugar production in southern Africa, and more specifically Malawi, South Africa, and Mozambique.

4 Countries and Commodities: Sugar and Southern Africa

Examples of the ways that the EU shapes the international market for sugar have already been mentioned in this chapter. Sugar powerfully demonstrates the ways in which international trade rules and regulation affect agriculture in less developed countries. The Common Agricultural Policy (CAP) of the EU supports and benefits a relatively small number of sugar producers in Europe and the implications of this are an undermining of markets and opportunities for vast numbers of farmers and agricultural workers in less developed countries by depressing world sugar prices.

Seven countries in southern Africa have significant sugar production: South Africa, Angola, Mozambique, Malawi, Zambia, Zimbabwe and Swaziland (see Table 2).

Aside from arguably South Africa, none of these countries are currently reaching their production potential. We will examine some of these countries in more detail.

South Africa: Sugar, Smallholders and Export Markets

For a country like South Africa, which regularly produces a surplus of 1.4 million tonnes of sugar per annum, the impact of EU policies is significant. In South Africa 55,000 small-scale farmers produce sugar and depend on a reasonable world market price in order to subsist, but Europe's subsidised exports continue to depress the price, undermining their ability to trade effectively. In some respects, the structure of the South African sugar industry reflects some of the structural constraints that are legacies of Apartheid. Unlike the sugar industry in neighbouring southern African countries, the South African sugar industry is a complex mixture of small-scale sugar cane growers, more industrialised larger-scale growers and a heavily industrialised processing and marketing sector dominated by two main players (Illovo and Tongatt-Hulett). In terms of the potential of the sugar industry for poverty alleviation and development, an increasingly significant feature of the South African sugar industry is the size of the smallholder sector, which is thought to consist of between 51,000 and 53,000 registered members.[33,34] The vast majority of these smallholders farm on Tribal Authority Land and are women, and thus provide an important economic input into what are generally poorly developed and poor rural areas in South Africa. These smallholders supply a significant proportion of the sugar supplied to sugar mills in South Africa; on average, 85%.[33]

The sugar industry is regarded as a major contributor to the rural economies of the sugar cane growing areas of South Africa, namely KwaZulu-Natal, Mpumalanga and the Eastern Cape, generating an income of almost R5billion per annum from the sale of sugar and molasses.[33] The South African sugar industry employs some 104,000 people and forward and backward linkages to supporting industries, directly and indirectly, employ another 20,000 people in the sugar producing regions.[35] Given the widely accepted dependency ratio of 1:5, over half a million people are sustained by the sugar industry in three of the poorest, most rural and least developed provinces in South Africa.[36] There are strong institutional relationships between smallholders,

Table 2 *Sugar production, consumption and potential in selected southern African countries, 2002*[10]

	Production (tonnes)	*Consumption (tonnes)*	*Potential (tonnes)*
South Africa	2,631,000	1,303,000	2,700,000
Swaziland	526,000	204,000	650,000
Zimbabwe	677,000	346,000	800,000
Malawi	202,000	159,000	280,000
Zambia	186,000	82,000	250,000
Mozambique	44,000	100,000	300,000
Angola	31,000	102,000	250,000

processors and the state, reflecting the important role sugar is expected to play in agrarian development and poverty alleviation in the sugar growing regions. Land redistribution initiatives, a certain level of tariff protection, the creation of infrastructure and support services, and research and extension all interlock to drive the development of the sugar industry in South Africa.

The exportation of EU sugar at subsidised prices to third markets–particularly in the Middle East and North Africa–is removing the opportunity for South African producers to sell there. For example, in 2001 the EU exported 770,000 tonnes of white sugar to Algeria, 150,000 tonnes to Nigeria and 120,000 tonnes to Mauritania.[37] These markets, in particular, are important to South Africa, but further opportunities are lost in the face of EU dumping of sugar. South Africa effectively has to sell over half of its annual sugar production on world markets at artificially low prices. A World Bank study, commissioned in 1990, indicated that as far back as 1984, the EU sugar regime was costing South Africa as much as US$50million per annum. In current US$ terms, coupled with the even lower current world market prices, that figure is likely to be closer to US$150 million per annum.[38]

Finally, South Africa is not only constrained by EU sugar regime, it is also not party to some of the preferential access agreements in the EU market, such as the EU 'Everything But Arms' initiative and the African, Caribbean and Pacific initiative (ACP).[39] These initiatives, set up in the name of development, allow, often very limited, preferential market access to some of the poorest countries in the world. South Africa is excluded from these initiatives as it is not classified as a poor enough country in order to benefit, but to exclude South Africa is to overlook the tremendous rural poverty that exists there; South Africa remains a country of extremes, with enormous sinks of extreme rural poverty and deprivation and, certainly in the sugar producing regions, sugar provides an important buffer against deeper and broader poverty.[40] South Africa thus loses out in terms of the artificial lowering of world market prices, unfair competition in third markets, pressures on its domestic sweets and chocolates markets, and on top of that is excluded from the benefits–however marginal–of any preferential EU market access initiatives. It is clear that in South Africa, unfair trade places constraints on several sectors of its sugar industry, but in particular it constrains the continued development and expansion of a large cohort of small-scale producers, depriving them of the profits they would need to diversify and develop their farms and livelihoods.[41] This is a significant issue in the face of the government's attempts to develop a wealthier, more productive cohort of export-oriented African smallholders.

Malawi: Sugar, Trade Agreements and Uncertainty

Very different from South Africa in the regional context, Malawi, Mozambique, Swaziland and Zambia are some of the lowest cost producers of sugar in the world. Mozambique and Swaziland are actively increasing sugar production. Indeed in Swaziland sugar is the dominant sector, accounting for 53% of total agricultural output.[42] Swaziland is planning to increase small-scale production to as much as 700,000 tonnes by 2008, an increase of over 30% on current levels. Currently, it is estimated that 80,000 people directly and indirectly gain employment as a result of sugar production in Swaziland. However, the depression of world sugar prices due to the EU sugar regime erodes the profitability and sustainability of sugar production in these countries. In Malawi, the sugar sector accounts for just 3.5% of GDP.[38] Current production levels average around 202,000 tonnes per annum and annual consumption averages 159,000 tonnes. Malawi is one of the six lowest cost sugar producers in the world and that is driving an expansion in sugar production towards its 280,000 tonne potential.[43] The Malawian sugar industry is almost totally controlled by one company, Illovo Sugar. Illovo has started expanding sugar production in Malawi and is supporting the development of smallholder schemes, which accounted for fully 10% of production during the 2001/02 season. The continued growth of these schemes, in the context of increased exportation of Malawian sugar, could be an important source of secure livelihoods in the future.

Unlike South Africa, as an officially designated 'less developed country,' in 2001 Malawi enjoyed a 23,000 tonne quota under the Sugar Protocol and an average of a 14,000 tonne quota under the Special Preferential Sugar arrangement. This has been equivalent to around 18.3% of Malawian sugar production.[43] In 1997/98, the estimated income transfers from Malawi under the preferential sugar arrangement was slightly over US$10million, this value was expected to rise to over US$20million by 2002/03. Whilst enjoying significant benefits through preferential sugar arrangements with the EU, Malawi is gaining access to significantly smaller income transfers than it might if EU quotas were larger, or tariffs were removed.[43]

The phasing in of the 'Everything But Arms' (EBA) initiative to replace existing agreements such as the Special Preferential Sugar arrangement means that current trade arrangements with the EU will be re-orientated.[44] The shifting nature of bilateral trade agreements is an issue for the Malawian sugar industry. Commerce and Industry Minister, Peter Kaleso stated in May 2001 that 'Our sugar market in Europe must be protected as part of the long term ACP-EU partnership… ACP countries, including Malawi, have requested the EU to legally protect our market in Europe.'[45] At issue was the benefit Malawi will accrue from the EBA initiative as opposed to the current trade arrangements.[46] Unlike the sugar protocol, EBA access arrangements provided no guarantee that the internal EU sugar price will be paid for its imported sugar.[47] There are potentially huge losses to be made for a country like Malawi, depending on how the EU chooses to organise the institutional arrangements that will determine EBA sugar import prices.[48] Compounding increasing uncertainty about the trading relationship between the EU and Malawi is increasing protectionism between southern African markets. Other regional sugar-producing countries such as Mauritius, Tanzania and Zimbabwe have been engaging in a round of tense trade negotiations, which have involved occasional internal subsidisation and banning of sugar imports as countries seek to protect their internal markets in the face of regional economic instability. Analysts fear that, despite Malawi being such a cheap producer of sugar, the industry will not survive if the health of the domestic market remains uncertain: a direct result of the structure and subsidies of the international sugar market.[49]

It is clear that in Malawi there has been a move towards supporting the creation of a population of sugar-producing smallholder farmers, or outgrowers, linked into contractual agreements with the key processor, Illovo. This can generate significant, if uncertain, income in rural areas, but again it is important to reiterate the constraint on Malawi's sugar production, operating at only two-thirds of its potential, and the uncertainty created by the sugar industry being so dependent on shifts within the EU's portfolio of sugar rules and regulations.

Mozambique: Regenerating a Sugar Industry

Besides Malawi, several southern African countries have a history of some sort of preferential trade arrangement with the EU, either through the EU Sugar Protocol or the EU Special Preferential Sugar scheme, or through the EBA agreement. However, this preferential trade invariably accounts for only a small percentage of total production.[1] The significant potential within southern Africa's sugar industries is in part constrained by the small scope of these preferential trade agreements. Most sugar must be sold on the unfettered international market. Mozambique, in particular, holds an enormous potential to increase its sugar production given the correct circumstances. As it stands, the sugar sector is already the largest source of employment in Mozambique, employing 23,000 workers as of 2001, one-third permanently and the rest seasonally. This number could increase to 40,000 if the sugar industry were to be successfully invested in, financed and rehabilitated.[50] In addition, forward and backward linkages with transport, packaging, storage services, marketing and intermediate goods production (such as molasses and bagasses for producing alcohol, cattle feed and paper) could create a further 8–10,000 jobs.[50] Furthermore, as sugar is less vulnerable to adverse climatic conditions than most other major crops in Mozambique, it is an important way of diversifying and stabilising household income. Environmental and social concerns about sugar cane production in Mozambique exist. Environmental concerns include stubble burning and the use of chemical pesticides, but soil degradation and erosion is low compared to other crops. Working on sugar plantations is physically strenuous but jobs are highly valued by rural households in Mozambique for the stable cash incomes they offer, enabling households to send their children to school, purchase clothing, invest in other agricultural practices and altogether improve their standard of living.

Mozambique is one of the countries to suffer most from the EU's sugar policies. Despite a stellar economic growth rate in African terms during the post-war period, Mozambique remains one of the poorest countries in the world. During the civil war many people lost their means to a living, infrastructure was destroyed and the agricultural sector deeply affected. Against this backdrop Mozambique maintains a primarily rural population, which continues to suffer broad and deep poverty. Almost 70% of people live below the income poverty line and agriculture continues to be the only source of employment and income for many.[50] Hence the rehabilitation of the country's sugar industry has become a priority in recent years, both as a source of employment and as a means of generating foreign exchange through the exportation of sugar. In an ideal world, Mozambique is ideally positioned to export. It is one of the lowest cost producers in the world, alongside Zambia and Malawi. At full capacity, production costs for refined sugar are just under US$280 per tonne–far less than one-half of EU costs at US$660 per tonne. Analysts suggest that rehabilitating and investing in the industry has the potential to lower the cost of production to US$180–235 per tonne.[51]

Attempts to rehabilitate the industry have faced many barriers.[52] Having essentially been blocked out of the EU market until 2002 when certain small trade agreements were negotiated, Mozambique has been unable to take advantage of the EU's high internal prices. The dumping of European surpluses on the world market, and the dominance of the EU in third country markets, have further depressed export prospects.[10] In addition, sugar imports from neighbouring countries–due to tariff loopholes in Swaziland and the chronic economic crisis in Zimbabwe–have continued to undermine domestic market demand. These conditions, which constrain the potential for exports and the potential of the domestic market, are creating a difficult environment in which the Mozambican sugar industry must operate.

These are not the only problems the sugar industry in Mozambique faces. The World Bank and International Monetary Fund (IMF) have been pressuring developing countries to cut their sugar tariffs. The World Bank and IMF maintain that they give the same advice on tariff cutting to both rich and poor countries alike. The key difference is that richer countries are not bound to and dependent on the loans of the international financial institutions and can, and do, ignore their advice when it is in their interests to do so. Not so for developing countries such as Mozambique.[50,53,54] This bifurcated arrangement, where richer and poorer countries have the political and economic wherewithal to react differently to the advice of international institutes such as the World Bank and IMF, gives rise to much of the anger and frustration directed towards the WTO. Again, the notion of one form and trajectory of globalisation and the ways in which nation states react to it does not appear to fit the reality in countries such as Mozambique.

Mozambique is one of the lowest cost sugar producers in the world, but despite this the enormous subsidies given to EU sugar producers means that, like Malawi, it is vulnerable to cheap sugar imports. Mozambique has been put under increasing pressure from the IMF to remove its sugar import tariffs–a move which would wipe out the Mozambican sugar industry in its entirety. In 1999, the IMF tried to persuade Mozambique to become the first country to withdraw protection. The World Bank has chosen to proffer a similar set of recommendations. The position of the government of Mozambique is clear. They argued a degree of protection is necessary in the face of the decline in the international price of sugar as the unsubsidised domestic market could not compete with the potential import of subsidised sugar sources from more developed countries.

Mozambique has had a long and fractious history of sugar production. Historically, the 19th-century sugar industry was built on the coercion of a labour force by colonial authorities. British investment followed, and by 1972 output had reached 325,000 tonnes. During the liberation war, forced labour ended and sugar estates began building infrastructure for their employees.[55] Independence led to the abandonment of five of the six plantations, the running of which was taken over by the state. Post-independence, the governing party, FRELIMO, gave high priority to workers and working conditions and took over the running of the plantations.[56]

Structurally, large plantations, each with a sugar mill, dominate the sector.[50] More recently, private companies have reluctantly taken on increased responsibilities for housing, health and education provision as part of their social contract with their workers. The government has been seeking to spend US$250million on rehabilitating the sector.[50] Despite this, working conditions are not ideal. Pesticide sprayers do not always have the correct protective clothing, people involved in cutting, too, are afforded little protection. Hours are long and arduous, and conditions are hot. Malarial infection is also a common problem.[57] Little regulation of labour conditions exists. Despite the poor working conditions, employment in the sugar industry is highly prized in Mozambique. Unskilled, often seasonal, workers earn around US$30 per month and this can play an important role in securing sustainable livelihoods for people. Research suggests that a steady cash income, even if it is relatively small in dollar terms, can play an important role in building the assets of the rural poor.[58] In particular, without a steady cash income rural households tend to become increasingly vulnerable and risk averse, and this has considerable knock-on effects for other agricultural practices and sectors, including subsistence agriculture.[59] Cash flowing into rural areas serves to lubricate and support the range of productive activities that households and communities undertake. This is particularly important in the context of Mozambique where the vast majority of economic growth and development assistance has accrued in the south, to the capital, Maputo, and its environs. Mozambican rural households tend to rely on a mix of agriculture for food and sale, wage labour and informal trade. The average household is dependent on an injection of some cash income.[60] The sugar estates tend to be in areas short of employment opportunities and therefore further aid local economies. Increasing the scale of the Mozambican sugar industry would appear to have far-reaching consequences for small-scale agriculture and rural livelihoods in the areas surrounding the sugar estates. In this way, commercial export agriculture provides capital to invest in subsistence agriculture and provides a bulwark against future stresses and challenges.

Mozambique represents a different set of impacts wrought by unfair trade in comparison to South Africa and Malawi. Large, essentially agro-industrial sugar plantations, that play important roles in rural economies, are constrained from reaching full production capacity because of the decline of the international market price of sugar, uncertainties surrounding accessing EU markets and the threat of imports of cheaper regionally produced sugar. Mozambique is a country that, despite enormous economic growth and development assistance, has struggled to develop its rural hinterlands. Investment in the revitalisation of the sugar industry appears to be one pathway to regenerating rural livelihoods. Again, it appears the EU sugar regime curtails the degree of progress that can potentially be made; it has been estimated that if Mozambique had unrestricted access to EU sugar markets, it could have earned up to US$38million more in 2004.[10]

5 Conclusions: Re-Connecting the Global and the Local

The case studies of the sugar industry in Malawi, South Africa and Mozambique illustrate the far-reaching consequences of unfair trade on agriculture in less developed countries. Just as it is important to remember there is not just one trajectory of globalisation, there is not one type of agricultural system, nor one type of less developed country. However, it is evident that unfair trade practices impact upon smallholder sugar producers, agro-industrial producers and local sugar processors in important constraining ways. There are further knock-on impacts within local communities and rural economies. Rural livelihoods and southern African agricultural systems are complex and diverse, evolving within their particular agro-environmental contexts and within the rules and regulations that support or inhibit the choices and opportunities available to them. Accordingly, rural livelihoods have developed to cope to a certain extent with the vagaries and inequities of international trade, through diversification of activities, seeking urban wage labour, or the exploitation of their own labour or resources. Agricultural systems and livelihoods are evolving in the context of constraint wrought by unfair international trade. This does not mean, however, that a reorientation of the way international trade is conducted does not present enormous opportunities for profit, investment and growth within the sugar sector, amongst others, in the developing world.

Indeed, the reflexive way in which rural life is conducted in southern Africa indicates that changes in the EU sugar rules and regulations will quickly be taken advantage of, and change exists in the not-too-distant future. Disputes between the US and the EU strain the relationships between the great northern trading blocs, the increase in the size of the EU makes CAP even more unsustainable, the continuing development of the agro-food sector raises uncertainties, and the 'group of 23' developing countries nations who acted as a bloc at the 2003 WTO Ministerial Meeting in Cancún illustrated the ways alliances can be turned in the favour of less powerful countries. These global disputes, politicking and new alliances, are again very far removed from the realities of a smallholder sugar producer in deep rural Malawi or South Africa, or a labourer working on a sugar estate in Mozambique. The reality is that the intangibles of politics and pressures that may recast the way global international trade is conducted will never touch on the rural areas in less developed countries where unfair trade hits hardest. This global–local political disconnection does not render the many economic connections, some of which have been highlighted in this article, any less tangible. In fact, in many ways perhaps it makes them more immediate.

References

1. Oxfam, *Rigged Rules and Double Standards: Trade, Globalisation, and the Fight Against Poverty*, Oxfam International, 2002.
2. P. Gibbon, Present Day Capitalism, the New International Trade Regime and Africa, *Review of African Political Economy*, 2003, **29**, 91.
3. P. Richards, *Indigenous Agricultural Revolution: Ecology and Food Production in West Africa*, Hutchison, London, 1985.
4. P. Blaikie, *The Political Economy of Soil Erosion in Developing Countries*, Longman, London, 1985.
5. D. Dollar and A. Kraay, Growth in Good for the Poor, *Journal of Economic Growth*, 2002, **7**, 3.
6. B. Milanovic, The Two Faces of Globalisation: Against Globalisation as we Know it, *World Development*, 2003, **31**, 4.
7. M. Law and C. Barnett, After globalisation, *Environment and Planning D*, 2000, **18**, 1.
8. A. Escobar, *Encountering Development: The Making and Unmaking of the Third World*, Princeton University Press, Princeton, 1996.

9. International Sugar Organisation, *Sugar Year Book*, ISO, London 2002.
10. Oxfam, *Dumping on the World: How EU Sugar Subsidies Hurt Poor Countries*, Oxfam GB, 2004.
11. The export subsidy is so large that for every €1 generated through export sales the EU have to pay out subsidies to the tune of €3.30.
12. Court of Auditors, Special Report No. 20/2000 Concerning the Management of the Common Organisation of the Market for Sugar, Together with the Commissioner's Replies, *Official Journal of the European Communities*, 2001/**C50**/01.
13. Calculations for Oxfam based on the profits and loss accounts of British Sugar and Associated British Foods for 2001. *Bureau van Dijk AMADEUS* database.
14. A 'third market' in this case refers to a country outside of the EU and to which African states may try to export.
15. World exports were 16.25 million tonnes in 2000/01. EU exports were 4 million tonnes (expressed as tonnes of raw sugar). Source: FO Licht. International Sugar and Sweetener Report, **34**.
16. Oxfam calculation. Based on estimate of production capacity by Joseph Hanlon and personal communication with Patricia Jamieson of Tate and Lyle on EBA prices.
17. M. Goodman and M. *Watts in Globalising Food: Agrarian Questions and Global Restructuring*, M. Goodman and M. Watts (eds.), Routledge, London, 1997.
18. H. Friedman, The Political Economy of Food, *New Left Review*, 1993, 197.
19. See Mozambique, in Section 4 of this chapter.
20. Defining exactly what a rural livelihood is proves problematic. The diverse nature of livelihoods means that any workable definition must be broad. One of the most widely used and enduring is Chambers and Conway's definition: A livelihood comprises the capabilities, assets (including both material and social resources) and activities required to construct a means of living. R. Chambers and G. Conway, *Sustainable Rural Livelihoods: Practical Concepts for the 21st Century*, IDS Discussion Paper 296, IDS, Sussex, 1992.
21. J. Pretty, *Regenerating Agriculture*, Intermediate Technology Publications, London, 1995.
22. I. Scoones, C. Chibundu, S. Chikera, P. Jeranyama, D. Machaka, W. Machanja, B. Mavedzenge, B. Mombeshora, M. Mudhara, C. Mudziwo, F. Murimbarimba and B. Zirereza, *Hazards and Opportunities: Farming and Livelihoods in Dryland Africa*, Lessons from Zimbabwe, Zed Books, London, 1996.
23. S. Devereux, Goats Before Ploughs: Dilemmas of Household Response Sequencing During Food Shortages, *International Development Studies Bulletin*, 1993, 24.
24. J. Heyer, Landless Agricultural Labourers' Asset Strategies, *IDS Bulletin*, 1989, **20**, 2.
25. P. Richards, in *Farmer First: Farmer Innovation and Agricultural Research*, R. Chambers, A. Pacey and L. A. Thrupp (eds.), Intermediate Technology Publications, Sussex, 1989, 40.
26. A. De Waal, *Famine that Kills: Darfur*, Sudan, 1984–1985, Clarendon Press, Oxford, 1989.

27. R. Baber, in *Land, Labour and Livelihoods in Rural South Africa*, Volume 2, M. Lipton, F. Ellis and M. Lipton (eds.), Indicator Press, Johannesburg, 1996.
28. C. Mather, Debating Rural Livelihoods in South Africa, Guest Review Article, *Canadian Journal of African Studies*, 1998, 18.
29. Interview data collected central Kenya, April–May 2004.
30. D. Bryceson, C. Kay and J. Mooij, Disappearing Peasantries? *Rural Labour in Africa, Asia and Latin America*, Intermediate Technology Publications, London, 2000.
31. H. A. Freeman, F. Ellis and E. Allison, Livelihoods and Rural Poverty Reduction in Kenya, *Development Policy Review*, 2004, **22**, 2.
32. Daily Mail and The Guardian, 3 December 1999, *Number of Farm Workers Declines*. Business Day, 18 October 2000, *New Data to Address Job Losses on Farms*.
33. M. Maloa, *Sugar Cane: A Case as Development Crop in South Africa*, paper presented at the SARPN conference on Land Reform and Poverty Alleviation in Southern Africa, 4–5 June, 2001.
34. Smallholder participation in sugar production varies annually with, in the 2000/02 season, only around 31,000 registered small growers delivering sugar to refineries.
35. Linkages with the South African chocolates and sweets industry are strong. This industry is also under threat as its domestic market share is being undermined by imports derived from subsidised sugar.
36. J. May, in *Land, Labour and Livelihoods in Rural South Africa*, Volume 2, M. Lipton, F. Ellis and M. Lipton (eds), Indicator Press, Johannesburg, 1996.
37. Oxfam, *The Great EU Sugar Scam: How Europe's sugar regime is devastating livelihoods in the developing world*, Oxfam, Oxford, 2002.
38. European Research Office, *Implications of the Reform of the EU Sugar Regime for Southern African Countries, Part 3: Country Profiles*, compiled by the ERO in collaboration with Oxfam GB, 2001.
39. Preferential access granted to the African, Caribbean and Pacific (ACP) countries only includes raw cane sugar. Refined sugar imports are tightly controlled, again limiting opportunities for value-adding in a country like Mozambique which already has refining capacity.
40. Statistics South Africa, *South Africa Survey, Millennium Edition*, Government Printers, Pretoria, 2000.
41. C. Mather and A. Adelzadeh, Macroeconomic Strategies, Agriculture and Rural Poverty in Post-Apartheid South Africa, *Africa Insight*, 1998, **28**,1.
42. United Nations Conference on Trade and Development, *Policies for Small-Scale Sugar Cane Growing in Swaziland*, UNCAT, Geneva, 2000.
43. ASSUC, *EBA–An Impact Assessment for the Sugar Sector*, Association des Organisations Professionnelles du Commerce des Sucres pour les Pays de l'Union Europèene, Brussels, 2001.
44. In 2000 the EU discussed launching an initiative to provide tariff and quota-free access to its market for all products except armaments from the 49 least developed countries–the 'Everything But Arms' Initiative (EBA). That included full access to the EU sugar market for LDCs. By the time the initiative was launched

in 2001, European vested interests had successfully lobbied to slash back the deal on sugar. Instead of open access, sugar producing LDCs have been granted very restricted quotas that will be incrementally increased.

45. Malawi pushes for sugar export protection, *The Nation*, 2 May 2001, Lilongwe, Malawi.
46. European Research Office, *Implications of the Reform of the EU Sugar Regime for Southern African Countries, Part 2: Reform of the EU Sugar Regime: Issues Arising in EU–Southern Africa Sugar Sector Relations*, Compiled by the ERO in collaboration with Oxfam UK, 2001.
47. This is a crucial caveat, as internal EU sugar prices are generally two to three times higher than world market prices.
48. Potentially at issue is the monopolistic structure of the major European sugar refiners, who could be in a very strong bargaining position regarding contract negotiations.
49. The Nation, *Illovo plays down pressure from Zim sugar*, Lilongwe and Blantyre, 11 March 2003.
50. Instituto Nacional Do Açúcar, The *Sugar Sector in Mozambique: Current Situation and Future Prospects*, Ministry of Agriculture and Rural Development, Mozambique, 2000.
51. Netherlands Economic Institute, *Evaluation of the Common Organisation of the Markets in the Sugar Sector*, Netherlands Economic Institute, Rotterdam. 2000.
52. P. Robbins, *Review of the Impact of Globalisation on the Agricultural Sectors and Rural Communities of the ACP Countries*, Study for the Technical Centre for Agricultural and Rural Cooperation, **CTA no. 8007**, 1999.
53. International Monetary Fund Staff Report on Mozambique, 17 January 2001.
54. World Bank, *Mozambique. Country Economic Memorandum: Growth Prospects and Reform Agenda*, **Report No. 20601-MZ**, 2001, 9. See also ref. 51, p.11.
55. J. Hanlon, *Mozambique: the Revolution Under Fire*, Zed Books, London, 1984.
56. Like many immediate post-colonial African states, FRELIMO was based on socialist principles.
57. T. Wandschneider and J. Garrido-Mirapeix, *Cash Cropping in Mozambique: Evolution and Prospects*, European Commission, 1999.
58. D. Bryceson, De-agrarianisation and rural employment in Sub-Saharan Africa: a sectoral perspective, *World Development*, 1996, **24**, 1.
59. C. Vogel and J. Smith, in *Global Desertification: Do Humans Cause Deserts*? J. Reynolds and D. Stafford-Smith (eds.), Dahlem University Press, Berlin, 2003.
60. C. Boyd, J. Pereira and J. Zaremba, *Sustainable livelihoods in southern Africa: Institutions, Governance and Policy Processes: Mozambique Country Paper*, SLSA Paper, 2000.

Free Trade in Food: Moral and Physical Hazards

COLIN D. BUTLER

1 Introduction

The essential principle of 'free' trade is the abolition of artificial subsidies supplied in the production of goods and services, and the absence of tariffs (customs duties) imposed upon the consumers of the same goods and services. The most famous classical economist associated with free trade is the Englishman David Ricardo (1772–1823). Ricardo claimed that free trade, operating through the principle of 'comparative advantage' can benefit producers *and* consumers, creating a competition in which no-one is a loser.[1] Simply stated, this principle proposes that producers should specialise in the growing or production of goods and services for which the supplier is best suited. For example, tropical countries will have an advantage in the growing of bananas, compared with countries at high latitudes. Comparative advantage can be determined by more than climate. Regions with a long tradition of specialisation in a craft, industry or agricultural skill are likely to be more efficient producers than a novice population. Ricardo's genius was to show that specialisation of goods produced using comparative advantage could avoid 'zero sum limits'.[2] That is, free trade between parties each specialising in their greatest comparative advantage was claimed to result in a game where all competitors could in fact be winners–a form of free lunch.

At times, advocates of trade liberalisation have claimed that a wider adoption of free trade is an inevitable step in an inexorable march of history towards global prosperity. For example, Margaret Thatcher's proclamation 'there is no alternative' (to market deregulation–a synonym for free trade) achieved notoriety as the acronym TINA. Typical of the view in support of free trade is the following: 'We know from David Ricardo that a country enjoys higher welfare if, instead of directly producing all the goods its members wish to consume, it specialises in the production of goods

Issues in Environmental Science and Technology, No. 21
Sustainability in Agriculture

it can produce at comparatively lower cost, and exchanges the surplus for those it does not produce in sufficient quantity.'[3]

Unlike most free lunches, free trade *can* avoid some zero sum limits. Greater efficiencies, enabled by comparative advantage, are often possible. Nevertheless, free trade, as practised in the real world, has many hidden costs. In most cases, one diner, the proposer, enjoys a bargaining position far superior to his or her partner. As well, the chef and the waiter who prepare and serve the meal are likely to be underpaid. Sometimes their labour is guaranteed by coercion. Finally, the food consumed at this lunch is often invisibly tainted, not only by odourless and tasteless poisons, but also by many 'shadow' costs, including environmental degradation that may only become apparent in the distant future.

Like many fables, this critique of free trade is an oversimplification. Sometimes even substantial environmental damage may be justified if it lessens human misery. But an unqualified endorsement of free trade is unjustified, especially from the position of the weaker party to the negotiation. This chapter starts with a critique of comparative advantage, arguing that its benefits have been overstated. It then discusses four main kinds of economic *externalities* uncounted in the cost of food. These are (a) the hardships, risks and exploitation endured by many human beings, both directly and indirectly engaged in the production of food for export, (b) the cruelty to many animals raised in 'factory farms', (c) the contamination of food by infectious agents and toxic wastes and, finally, (d) the damage to the wider environment, especially to the climate system and to ecosystems. The chapter does not argue that these problems can be entirely eliminated, but points out that the cost of these moral and physical hazards is immense. These hazards should, and can, be lessened.

A more complete accounting of these hazards would create financial and transaction costs that would increase the cost of food to the consumer. The payment of higher wages to food exporters in developing countries will reduce global inequality. For both reasons, reforms are likely to be resisted. Nevertheless, the article explores some practical ways to reduce these moral and physical hazards.

2 Free Trade – Theory and Reality

Ricardo's argument that comparative advantage could escape zero sum limits is illustrated in the following example. Assume, in pre-trading England and Portugal, that equally skilled, healthy and remunerated workforces exist, each of ten people. Assume that these workers constitute the entire economy, and their wages are paid, in kind, in proportion to their output. In each country, we assume that half of this workforce is engaged in producing each commodity, but that the average clothmaker in England is twice as productive as his counterpart in Portugal, and that the reverse is true for making wine. Using these assumptions, each country produces 45 units of cloth and wine over 30 days (see Table 1).

Ricardo's genius was to demonstrate the consequences if each country specialised in producing the goods for which it had the greatest *comparative advantage*. Ricardo noted that different regions, countries and workforces were more or less efficient producers, because of factors such as geography, soil, climate, culture, history and chance. In the example under consideration, Portugal has a comparative advantage in producing

Table 1 *Pre-specialisation. Both countries produce each good. Total production during 600 person-days of labour is 90 units for the two countries*

	Good	*Person-days/Unit*	*Workforce*	*Person-days Worked*	*Production (30 days)*	*Total*
England	Cloth	5	5	150	30	
	Wine	10	5	150	15	45
Portugal	Cloth	10	5	150	15	
	Wine	5	5	150	30	45
Total			20	600		90

Table 2 *Post-specialisation. Each country specialises in the production of a single good. Total production increases to 120 units, yet no more person-hours are worked*

	Good	*Person-days/Unit*	*Workforce*	*Person-days Worked*	*Production (30 days)*	*Total*
England	Cloth	5	10	300	60	
	Wine	10	0	0	0	60
Portugal	Cloth	5	0	0	0	
	Wine	10	10	300	60	60
Total			20	600		120

wine, but England is more suited to making cloth. Specialisation allows the same sized labour force to produce more goods in the same time. This is shown in Table 2.

In this case, specialisation boosts the total production of wine and cloth, without engaging any more labour. Despite additional costs, such as transport, it seems plausible that these costs will be outweighed by greater production. This is true even if other costs are considered, such as the dismantling of cloth-making machinery in Portugal, and producing labels in a different language. These gains are maximised if no country applies a tariff to the imported goods.

The efficiencies used in this example may be overstated. The extent of comparative advantage may change over time, for example as a workforce gains more experience or becomes more capitalised. A closer examination of Ricardo's theory reveals additional problems. For one thing, it assumes that there are no diminishing returns of labour and land productivity. That is to say, while Portugal may have a clear advantage over England in making the first million litres of wine, production of the second million litres may rely on land, labour or climatic conditions that are not as well-suited, reducing the gains.

Another issue is the problem of monopolies. In this example, following specialisation, England will enjoy a monopoly on cloth, and Portugal will have a monopoly on wine. Either party could abuse this position, especially if there is asymmetry in either the number of monopolies held by one country, or asymmetry in the demand for the goods produced by that monopoly.

In a perfectly fair world (and assuming no significant transport or transaction costs), the extra productivity could translate to higher real wages in both countries (see Table 3). In the real world, wages and prices are influenced by the relative economic power of the broader economy within which any workforce is located, as well as by supply and demand. Consider if a third country (*e.g.* Spain) also has a comparative advantage over Britain in the production of wine (see Table 3). After specialisation, the supply of wine is then is likely to exceed demand, consequently depressing the unit price of wine. This can then lower wages, in both Spain and Portugal. England, as the sole producer of cloth is then in position to arbitrarily increase the price of cloth, and this could further increase the disparity in the wages paid in the three countries.

These are simplified examples, for example by assuming that the unit of production is a country rather than a firm. Nevertheless, they demonstrate that free trade and comparative advantage do not necessarily result in gains for *all* parties, and that increases in absolute and relative inequality are possible. Real wages of commodity producers will tend to fall if supply exceeds demand, even if producers exercising their comparative advantage generate increased total productivity. Specialisation can make nations vulnerable to oversupply, gluts and reduced demand for its exports, because of substitution or a change in fashion. It also means that a country can seal in a long-lasting or even permanent comparative disadvantage that might otherwise only be temporary. These drawbacks are far from theoretical.

Table 3 *Before specialisation, wages in each country are equal at 4.5 per person. After specialisation, the price of wine falls, because of its over-supply. Even though the wine-specialising countries have increased their productivity, wages decline relatively, and may even decline absolutely, depending on the elasticity of demand for wine*

	Cloth			*Wine*			
	Production	*Demand*	*Price*	*Production*	*Demand*	*Price*	*Wages*
England	30	20		15	25		4.5
Portugal	15	20		30	25		4.5
Spain	15	20		30	25		4.5
Total	60	60	1	75	75	1	
				pre-specialisation			
England	60	20			25		6
Portugal		20		60	25		3.75
Spain		20		60	25		3.75
Total	60	60	1	120	75	0.625	
				post-specialisation, unchanged wine demand			
England	60	20			28		6
Portugal		20		60	28		4.2
Spain		20		60	28		4.2
Total	60	60	1	120	84	0.7	
				post-specialisation, increased wine demand			

Problems with Ricardo's idealised theory were pointed out in the 19th century by the Austrian economist Friedrich List.[4] Yet his criticisms still remain largely unknown, both to supporters and opponents of free trade. Instead, the alleged virtues of free trade are repeatedly proclaimed as an important tool of benefit to both supplier and consumer. This is a reassuring myth in a vastly unequal world. The myth hints that economically powerful nations actually have the best interest of their weak trading partners at heart, and that, somehow, the continued poverty of Third World nations is despite the efforts of rich countries to give them an equal opportunity. No doubt this myth helps to assuage the conscience of policy makers, economists, consumers and others who benefit from decades of exploitation masked as fairness. Giving equal attention to List's critique would risk the unveiling of this conceit, and suggest that modern economic theory is not as benign as its advocates claim.

In recent decades, many countries have reduced their levels of protection, especially for manufactured goods. The low wage, well disciplined labour force in many developing countries, allied with increasingly sophisticated management systems and the use of advanced technologies, has seen significant increases in the export of manufactured goods to developed countries, resulting in the 'offshoring' of a significant number of jobs. The 1970 Nobel Laureate in Economics Paul Samuelson, one of the grand old men of modern economics, has dared to criticise free trade theory.[5] However, Samuelson's critique does not concern the hazards discussed in this chapter. Instead, he speculates that the offshoring of jobs from developed countries has reached such a scale that its benefits to consumers threaten to be outweighed by damage to the social fabric and living standards of working class communities in developed countries. Samuelson says little about the gains to workers in developing countries, but he implicitly supports the 'race to the bottom' critique which argues that a small class of privileged consumers stands to gain, as long as production costs fall, irrespective of their geography. That is, Samuelson points out that these gains in fact come at a price to other people.

While many developed countries have substantially reduced tariffs and subsidies on manufactured items, they have been far more cautious in commensurate subsidy reductions for agricultural produce. But many developing countries, especially in the tropics, with comparatively small areas and poorly educated labour forces, have accepted (or have been virtually forced, such as Ghana was shortly after its independence) the virtues of comparative advantage. More than 50 such developing countries depend on three or fewer primary commodities for more than half of their export earnings.[6] These countries are highly vulnerable to the decline in the price of many food commodities that has occurred since 1980 (see Table 4). Critics of the free trade in food have observed 'The potential for profitable agricultural expansion lies not in feeding the hungry but in better serving the markets of those with plenty to spend.'[7] Similarly the 'extraction and relaying of surpluses' has been identified as a central function of the poor, in serving wealthier people in both developing and developed countries.[8] A Filipino shrimp farmer lamented:

> 'The shrimp live better than we do. They have electricity, but we don't. The shrimp have clean water, but we don't. The shrimp have lots of food, but we are hungry.'[9]

Table 4 *The price of many important commodities declined in real terms between 1980 and 2000.*[6]

Decrease by 0–25%		*Decrease by 25–50%*		*Decrease by more than 50%*	
Bananas	4%	Coconut Oil	44%	Cocoa	71%
Tea	8%	Fishmeal	32%	Coffee	64%
		Groundnut Oil	31%	Palm Oil	56%
		Maize	42%	Rice	61%
		Soybean	39%	Sugar	77%
		Wheat	45%		

Coffee

Second only to oil, coffee is the world's most heavily traded commodity. It has long been overproduced. Most producing countries perceive advantage in the production of *more* rather than *less* coffee because any country that refrains from production will lose market share. Farmers committed to coffee are naturally reluctant to abandon their investment. Coffee growers, mostly in developing countries, also have a relatively compromised position with regard to information. The consequence of coffee overproduction is a depressed price. This suits consumers, not least because the wide geographic dispersion of coffee plantations reduces the vulnerability to supply from adverse weather or crop diseases which are unlikely to be global in scale.

A successful coffee cartel is unlikely. Unlike oil, coffee is produced by many countries, and, also unlike oil, coffee decays after production. Oil suppliers that temporarily refrain from production bear little if any long-term penalty, unlike coffee growers that refrain from harvest for one year. To be successful, a coffee cartel would require a governing mechanism, able to plan, enforce and compensate growers that refrain from production in many countries, which is an unlikely prospect.

In 2002–03, global production of coffee was 117 million 60 kg bags, but of these only 108 million were consumed. Overproduction resulted in historically low coffee prices (adjusted for inflation) of about US$1.10 per kg. Of the total retail value for 2002–03 of US$70 billion, producing countries received less than US$6 billion. This is a significant decline compared to the early 1990s, when producer countries earned about US$10–12 billion from a total of US$30 billion retail value.[10] Most of the value of coffee is retained in the largely wealthy countries that buy coffee, accruing to middlemen and retailers.

Tasmanian Potatoes

Problems from competition among food suppliers are not confined to developing countries. The island state of Tasmania grows about 600,000 tonnes per annum of chipping potatoes, supplying about two-thirds of the Australian market. Until recently a handful of buyers have held the price at A$200 per tonne at the factory gate. The same tonne of potatoes (with oil) retails for up to A$9,000 if sold by the serving at McDonalds. Of course, additional labour and other costs are involved in the transportation from farm gate to mouth, but a difference of this size is indicative of exploitation. Eventually, the Tasmanian potato growers became sufficiently desperate and organised to win a price increase.[11]

These two examples show how free trade exploits competition between relatively numerous producers who are similarly poor with regard to information, capital, and economic and political power. Kenyan coffee growers can indeed grow coffee more easily and efficiently than in German greenhouses. But this has not brought them the prosperity suggested by the most naïve interpretations of Ricardo's theory.

3 The Moral Hazards of Free Trade

Any fall in commodity prices creates incentives to reduce production costs, including the cost of labour. Already-poor economies are perversely forced, through competition, to undercut the prices and living standards of producers in other poor countries. Where labour is plentiful, capital scarce, and the maintenance of machinery difficult, it makes economic sense to reduce the wages and conditions of as many workers as possible, thus lowering production costs and winning some exports.

These market forces mean that a substantial, though uncertain fraction of many commodities that are eaten, drunk and sipped are the products of Third World workers with few rights and few opportunities, particularly in comparison to First World consumers, who in most cases are unconscious of the scale of this divide. The 'fair trade' movement has publicised and reduced some of the most egregious forms of exploitation, and developed niche markets for ethically produced goods, such as coffee. It seeks to find markets for goods produced more ethically, particularly by the payment of higher wages. Consequently, fairly traded goods are more expensive. The global sales of the most widely fair-traded good, coffee, are less than 1% of global production, and the percentage of fair trade in other commodities is probably even lower.[12]

The Abuse of Human Beings

> 'Proper economic prices should be fixed not at the lowest possible level, but at a level sufficient to provide producers with proper nutritional and other standards.'
>
> *John Maynard Keynes* (1944)[13]

While the most extreme examples of inequality between consumer and producer are found in the international food trade, exploitation is also involved in the domestic growing of some labour-intensive First World crops. For example, the fruit and vegetable industry in the US has long relied on a comparatively poorly paid, pesticide-exposed, health-care deprived workforce, much of which is vulnerable, illegal and even desperate.

The problem of newly arrived immigrants from developing countries underpinning labour-intensive food production in developed countries is not restricted to the US. Recently, in the UK, more than 20 Chinese workers, thought to be vulnerable, illegal immigrants, drowned in treacherous Morecambe Bay while collecting valuable cockles.[14] Although some analysts claim that free trade does not depress working conditions,[3] market forces can be clearly seen at work in this case, depressing wages and safety for this dangerous work.

The issue of an underpaid and exploited workforce being used to produce goods in competition with a freer work force has long been an issue. In the US, the import of goods produced using convict labour was banned in 1890.[15] Goods produced using prison labour are still banned by the World Trade Organisation, but there is no exclusion for produce grown using child, slave or indentured labour. There are also claims that some goods sold on the world market *are* produced by prisoners, kept in Laogai (Chinese 'reform through labour' camps).[16]

The suffering and risk involved not only in the making of food but in the myriad of elements that support its production and global distribution are rarely included in the calculus of cost and benefit of trade, which instead mainly uses indicators such as the volume, price and monetary profit. This suffering and risk thus constitutes a moral hazard of free trade. In situations where fairly traded goods are either unavailable or considered too expensive, how can the consumer discriminate in order to avoid an unwitting endorsement of the most egregious exploitation? Currently, this is virtually impossible.

The Modern Economy: A Long Economic Chain

Central to ecology is the concept of 'food chains', in which the energy of a large number of simple organisms is consumed by a smaller group of organisms further up the chain, who in turn are vulnerable to consumption by better positioned predators. At the end of these food chains are found large and comparatively long-lived species such as eagles and sharks. Human beings now occupy the ends of most food chains, including those involving marine systems.

Less well-recognised is the parallel existence of a form of economic supply chain, in which the final human consumers benefit, usually indirectly, from the sweat, sacrifice and risk of the enormous number of people further down the economic chain. Our fellow humans have replaced non-human animals as our main competitor and predator, and some of our species ruthlessly exploit others through means such as debt bondage and slavery.[17] Particularly vulnerable to exploitation are groups perceived as 'other' by more powerful groups, whether of a different culture or language, or a lower class, caste, or 'claste'.[18]

Billions of people occupy positions in this chain which provide at least tolerable living conditions, with access to shelter, adequate nutrition, health care and a degree of freedom.[19] But near the start of this chain there exist many hundreds of millions of people who endure lives that few people nearer to the end of the chain could tolerate. The lives of these human beings are marked by risk, poverty, disease, fear and insecurity. One in six children of the world are child labourers.[20] At least eight hundred million are chronically hungry, and thus literally lack the energy to work themselves out of their poverty trap.[21] More than two billion people are estimated to be deprived of one or more essential micronutrients, especially of iron, iodine, vitamin A and zinc.[22] Most people with protein energy undernutrition probably also lack micronutrients; thus incurring a double burden, being deprived not only of energy, but also of cognitive potential, stolen by chronic micronutrient lack.[23]

Until recently, the world consisted of a myriad of separate economic chains. While some of these human systems were more equal than others, no society was ever entirely egalitarian.[24] Nevertheless, constraints, traditions and laws seem to have operated within most of these disparate chains to limit the degree of poverty and inequality.[25] In most cases the poor and wealthy lived in physical proximity and were more or less mutually visible. Principles of mutual obligation, religious duties and the possibility of open revolt usually restrained the worst exploitation. When inequality became visible and extreme, such as within Britain during the Industrial Revolution, many countering social forces arose which, over time, reduced the worst abuses.[26] These countering factors included the organisation of workers, sometimes supported by members of the upper class and intellectuals, to campaign for the vote and for better working conditions. As well, there is evidence that the government perceived that the deepening inequality could damage not only the social fabric but also national security. For example, though the declining height of the average British navy recruit, caused by chronic undernutrition, theoretically enabled smaller cabins, saving space and funds, it generated alarm because of the correlation of decreased height with reduced strength, stamina and fitness. In other societies, such as in Hindu India, a caste system evolved which, though imposing rigid controls and cruel conditions on members of the lowest caste, nevertheless provided limited rights to members of that caste, as well as generating a support system within it.

Today, the globalisation of trade has effectively melded thousands of economic systems into a single chain that embraces almost the entire global population. Probably the only people who survive outside this chain are a few indigenous populations, still clinging to subsistence lives, with minimal contact with the wider world. The number of these people is now miniscule, and their lack of acknowledged property rights makes them extremely vulnerable.[27]

While comparatively few people at the beginning of this global chain participate *directly* in food-exporting industries, virtually all consumers who live further along the chain benefit, albeit often indirectly, from the poverty of the poorest group. For example, assume that workers employed in a shrimp-processing factory in Chittagong, Bangladesh, which exports to Europe are paid reasonably well by Bangladeshi standards, so that they live above a level of absolute poverty (though far below the living standard of the average consumer of their produce in Europe). Because we have assumed that these workers are not subsistence workers, but participate in the Bangladeshi economy, it follows that the living standard of these workers will in turn rely on goods and services supplied, in the main, by workers who occupy an even lower economic level. In the context of many low-wage economies, the gap in living standards between those who constitute the first link of the chain and those further along who are employed in food-export industries may not be large, and may seem contextually reasonable. However, when viewed *across* economies the gap between many First World food consumers and many Third World food producers is profound and probably exceeds most historic gaps between rich and poor in either developed or developing countries. Thus all consumers in this global supply chain are both economically and morally connected to the poorest group.

The physical, linguistic and cultural gaps between consumers and producers disguise the economic gulf and weaken efforts, on either side, to narrow the gap. Although 'Fair Trade' makes an effort to improve the working and income conditions for many commodity producers, such as for coffee and cocoa, Fair Trade can do little to diminish this gap, or to improve conditions for people whose poverty underpins those involved in the export of crops, other than hoping that a 'trickle down' of money and values will gradually lead to reform.

Child Labour

Approximately 1.3 billion people globally (one in four) are employed in the agricultural labour force. A poor and vulnerable fraction of this group are landless labourers or subsistence farmers, who rarely, if ever, grow cash crops. But even poorer than this group are millions of child labourers, whose involvement has been documented in the production of many heavily traded food commodities including bananas,[28] cocoa,[29] oranges, shrimp,[9] and sugar,[30] tea and coffee.[31] Child labour may be as apparently benign as helping out after school on the family farm, but it can also involve relentless, heavy, ill-paid and dangerous work. For example, a recent report by the International Institute of Tropical Agriculture for the US Agency for International Development found that hundreds of thousands of children, most under 14 years, are engaged in hazardous tasks in the growing of cocoa, in the four West African nations (Ivory Coast, Cameroon, Ghana and Nigeria) that are the major source of this key ingredient of chocolate. These tasks including the spraying of pesticides, carrying heavy loads and using machetes.[32]

Similarly, a recent report in the New York Times claimed that Bonita branded bananas sold in the US depended in part on child labour.[33] Impoverished families are less likely to keep children at school, when they can be making even a small amount of money in exchange for hazardous conditions. Tasks undertaken by children include tying insecticide-laced cords between banana trunks.

In South-east Asia a substantial area of traditionally used land has, in comparatively recent history, been appropriated from indigenous populations, with minimal or no compensation, in order to grow palm oil, a hidden ingredient for margarine, chocolate, soups and potato crisps.[34] This appropriation of land is not a form of direct slavery or forced labour, but by depriving people of their ecological resources it has a similar effect.

The International Labour Organisation has estimated that 69,000 children of ages 10 to 14 and an additional 325,000 young people aged 15 to 19, were working in Ecuador (alone) in 1999. In circumstances similar to coffee, each 43 pound box of bananas purchased in Ecuador by exporters for $2 or $3 is sold for $25 in the US or Europe. Like coffee, the global oversupply of bananas for export forces the price lower.

The number, working conditions, life histories and life expectancy of the world's child labourers is poorly documented. Obstacles to such investigation range from employer and official non-cooperation to overt violence. Child labourers involved in agriculture are scattered in countless locations in dozens of countries. There are many advocates, especially in developing countries, for the use of child labour to produce traded goods, whether manufactured or agricultural.[15] These advocates argue that exclusion of goods produced by child labour would be a kind of protectionism, consigning labour-rich countries to even deeper poverty. This argument is plausible, not least because there is little evidence that those who propose boycotts against goods produced using child labour invest commensurate effort in finding an alternative livelihood for displaced workers.

There are compelling theoretical reasons for how untempered free trade can increase inequality, in a 'race to the bottom'. That is, competitive forces that operate in a market constructed only to reward the cheapest financial cost will inevitably reproduce and invent measures to lower the cost of production. This argument is especially strong for goods and services that do not require high levels of skill (because, by definition, highly skilled labour forces will be harder to exploit).

The use of child labour for unskilled agricultural work serves the short-term interest of both consumer and employer (that is, where employers compete against peers who also use child labour), not only by lowering the cost of labour, but also by reducing other costs entailed by employing an older and less compliant workforce, who are more conscious of exploitation, and also more knowledgeable of forms of escape from, or resistance to, exploitation.

In reality, the most extreme forms of such 'free' trade are limited, because even the most egregious markets and societies are tempered by the countering influences of international law and opprobrium. But, at the same time, the force of these countering winds is continually lessened by distance, blindness and greed. That is to say, when the area of production is far away, when the people involved in production are unknown, when the price to the consumer is low, and when the profits to the importer are great, the incentive to abuse human beings is high.

The Abuse of Animals

Countless animals are 'warehoused' in cruel conditions in 'factory farms'. In these landless farms most animals are treated as if they are inert factors of production, rather than sentient creatures. Urbanisation has greatly reduced the percentage of people with first-hand experience of farm animals. Factory farms are rarely visited, filmed or publicised. Consequently, few people know or seem to care about their conditions, even though these conditions would once have been regarded as abhorrent by many farmers. Examples of this abuse include calves that are reared away from the sun in order to keep their flesh pale, and of pigs kept in crates so tight that they cannot turn around or even lie down.[35]

The view that humans, like chimpanzees and other primates, evolved from forest-dwelling species was once widely ridiculed. An enormous gap was held to exist between the intelligence, and even the pain perception of humans and other animals. But this gap is increasingly under challenge.[36] At least some non-human animal species appear to be self-aware. There is also increasing evidence that many animals, from chickens to chimpanzees, are capable of complex communication, have distinct 'cultures',[37] extraordinary facial recognition, family bonding and even emotional reactions such as mourning and depression.[38,39]

We live in a global society that sanctions organ transplants for beloved pets,[36] yet forces billions of equally sentient creatures to crowded, housed lives with a complete absence of freedom. Defenders of intensively farmed animals point to regular feeding and the absence of predators (until the end) as compensatory virtues, but no doubt similar claims were once made about human slaves. To provide adequate nutrition for the human population of more than six billion probably requires a substantial degree of intensive animal farming, but the suffering involved is clearly an uncosted hazard of free trade. As with child labour, it should be possible to reduce this hazard, even though its complete elimination may seem utopian at present.

The Hazards of Intensive Animal Farming

Antibiotics and Hormones Many of the animals raised in intensive farms are fed antibiotics and hormones. These accelerate growth and reduce animal disease, but create hazards to human health, such as the inadvertent promotion of antibiotic-resistant organisms and the contamination of food with trace amounts of foreign chemicals, such as bovine somatotrophin.[40]

Nearly half of the total volume of antibiotics in the US is fed to animals, despite a strong scientific consensus that this is undesirable. Links exist between the sub-therapeutic use of antibiotics in animals and the increasing prevalence of resistant bacteria in humans. The World Health Organisation has advised against the practice of dosing animals with the same antibiotics used in human medicine. Yet the practice continues in many nations. For example, in 1996, the US Food and Drug Administration approved the use of the antibiotic class fluoroquinolines in chickens and turkeys, despite strong opposition from the Centers for Disease Control (CDC). Fluoroquinoline resistance quickly appeared in *Campylobacter* isolated from chickens, and by 1999 18% of *C. jejuni* and 30% of *C. coli* isolated from human patients showed fluoroquinoline resistance.[41]

Modern Herd and Flock Diseases Intensive farming practices have been directly linked with the emergence of several devastating herd and flock diseases, including 'mad cow disease' or bovine spongiform encephalopathy (BSE), foot and mouth disease (FMD) and avian influenza. The background of the BSE epidemic is well known. In order to improve the protein content in the diet of cattle, ground-up cattle remains, including the brain and spinal cord, were fed to cattle. This practice was claimed as economically 'rational' because it turned a waste product into a valuable food. But from an ecological perspective this practice was anything but rational, as cattle are normally vegetarian and certainly not cannibals.[42]

While it was originally argued that this practice is harmless, a similar disease called kuru, transmitted through the ritual cannibalism of human brains, was already known in New Guinea. In time, the causal agent of BSE, an unusual protein called a prion, was transmitted to humans, causing a devastating, rapidly progressive and still untreatable brain disease, called Creutzfeld–Jacob disease (CJD). So far, the size of this human epidemic has been modest, but transmission is probably still occurring, through blood transfusions[43] and surgical instruments that cannot be sterilised. As well as these human health effects, the BSE epidemic imposed an immense economic and psychological toll upon farmers, as millions of cattle had to be destroyed.

The genesis of the BSE epidemic is unsettling in the context of modern fish farming. A form of prion has recently been identified in fish.[44] Fishmeal is an important source of food for farmed carnivorous fish, and includes land animal as well as fish-derived protein. As well, under conditions of food scarcity, large, aggressive, transgenic, growth hormone-enhanced salmon cannibalise their smaller, non-genetically modified kin.[45] In sum, it is conceivable that neuro-degenerative diseases may one day be transmitted from eating farmed fish.

The 2003 epidemic of FMD in the UK also had a severe psychological impact on farmers. Unlike BSE, FMD does not pose a severe health threat to either cattle or humans, but it constitutes a significant *economic* risk. National herds that are free of FMD have an economic advantage; in order to maintain this advantage millions of cattle were again destroyed.[46,47]

The 2005 epidemic of bird influenza in several East Asian countries has led to the repeated culling of millions of birds. This flu has also been detected in pigs and cats, raising fears that genetic recombinants may enter the human population.[48] The crowded conditions in which many thousands of birds are housed facilitates the spread of these epidemics.

Food-borne Diseases Food is an important source of microbiologically transmitted disease, including gastroenteritis, hepatitis, peptic ulcer and, in some cases, gastric cancer. Gastroenteritis is a particularly important cause of child mortality in developing countries, but is also an important cause of morbidity in developed countries. On several occasions gastroenteritis in the US has been traced to food imported from developing countries, including Mexican strawberries and Guatemalan raspberries.[49] The most likely mechanism for the spread of these illnesses is that the food, designed to be eaten uncooked, was contaminated by sewage. The low standards of sanitation and hygiene in developing countries contributes to a lower production cost of these foods (than from fruit grown in a country with universal sanitation and clean water) but in some cases is accompanied by the invisible carriage of unwanted micro-organisms. Again, this hazard is not weighed in formal assessments of the gains of free trade.

4 The Physical Hazards of Free Trade

Even more subtle forms of food contamination occur than that due to micro-organisms, such as pesticide residues and mercury.[50] These costs are also not incorporated into calculating the cost of freely traded food.

Pesticide Residues in Crops

There is wide agreement that exposures to high doses of pesticides (likewise herbicides, fungicides and insecticides) is harmful to health. In many parts of the developing world, pesticide poisoning has become a favoured means of suicide.[51] In parts of Sri Lanka, more deaths are attributed to pesticide poisoning than to infectious diseases.[51] There is also broad agreement that the regulation, usage and education concerning pesticides in developing countries is poor.[52] Not all pesticides are equally harmful; many of the ones believed to entail the highest risk have been banned in developed countries, but are still used in developing countries.[53]

However, the relationship between *chronic* exposure to *low* levels of multiple pesticide residues, such as those absorbed through diet, and ill health is far more contested.[54] Advocates of pesticide safety can point to many epidemiological studies that have failed to find a substantial health risk from pesticides. However, critics of pesticide exposure, including within epidemiology, argue that existing epidemiological methods are a blunt instrument for detecting a causal effect from low-dose pesticide exposure and ill health. Critics point to four main limitations. These are (a) the difficulty of investigating the causation of diseases such as cancer and neurological conditions with development times as long as several decades, (b) the difficulty of measuring the cumulative dose of pesticide exposure, (c) the significance of interactions between different pesticides, and (d) the scarcity of adequate control groups. The fourth limitation may be the most important obstacle. That is, because exposure to multiple pesticides is now almost ubiquitous–within both developed and developing countries, it is extremely difficult to find otherwise comparable populations with little or no pesticide exposure. As well, there are few sources of funds to conduct proper studies, sometimes because of significant and powerful vested interests hostile to such research.

Nonetheless, a growing number of studies suggest that chronic exposure to low levels of pesticides may cause significant adverse health effects.[55] Farmers in developed countries have been repeatedly found to have higher rates of lymphoma than expected, possibly due to their exposure to multiple interacting pesticides.[56] Some groups appear genetically vulnerable to illness if exposed to pesticides. So far, this has been best documented for Parkinson's Disease,[57] but may also apply to breast and other forms of cancer. Children too are an important subgroup who may be more vulnerable to pesticide and other unwanted residues than other groups.[58]

Pesticides belonging to a class of chemicals called organochlorines (now banned from most developed countries) have been particularly suspected of harming health. This is considered plausible because of their propensity to bio-accumulate, their long half-life in human tissue, and, in some cases, their capacity to mimic oestrogen. In one particularly troubling study, exposure to the organochlorine pesticide dieldrin was assessed from blood samples collected from 7,712 women in 1976 whose health was then followed for 17 years. The investigators found that the level of dieldrin in 1976 predicted their subsequent development of breast cancer.[59] However, while other investigators have claimed that a causal link exists between breast cancer and pesticides[60] it is probably fair to say that this view is still provisional among most epidemiologists.

Whether or not pesticides are harmful, many consumers are unwittingly exposed to them, knowing nothing or little of the kind, dose or toxicity. Some of these consumers may be particularly vulnerable to adverse effects. This clearly represents a hazard of free trade. Which grower or exporter of food is likely to proclaim the unseen menu of pesticide and other residues in their product?

Chemical Residues in Fish and Marine Mammals

Residues of several organochlorines, including polychlorinated biphenyls (PCBs) and dieldrin have recently been detected in comparatively high concentrations in farmed, compared with wild, salmon.[61] Concentrations of chemicals among salmon farmed in Scottish waters were higher than among farmed fish in Washington state and Chile. Wild-caught fish from Alaska had, in some cases, concentrations 100 times lower than in the Scottish salmon. Just as in crops, debate continues about the health effects, if any, of these residues in salmon. But again, how many consumers can make an informed choice about the risk of exposure to these chemicals?

Another example of a food contaminant is the antibiotic chloramphenicol. This has been recently detected as a contaminant of Chinese honey. In susceptible people, this antibiotic can cause aplastic anaemia which is sometimes fatal. Honey contaminated by chloramphenicol is banned from sale in developed countries. Despite this, large quantities of contaminated honey continue to circulate on the world market.[62]

5 The Environmental Hazards of Free Trade

Climate and Ecosystem Change

The abundance of agricultural production available to affluent consumers in both developed and developing countries is subsidised by vast quantities of fossil fuels, burned and transformed by a handful of profligate generations, and by other forms of uncosted global environmental damage, including that to fisheries, forests and the atmosphere. As a result, an ever increasing number of humans are fed, but these environmental costs will be paid by future generations, who will inevitably have a reduced richness of possible experience and who may even experience compromised health and living standards as a result.

Food Miles

'Food miles' refers to the cumulative distance travelled and energy expended to assemble and distribute the components of a food and its packaging. A study from Germany in 1993 found that the ingredients in a glass jar of strawberry yoghurt travelled an average of 3,500 km.[63] Air transport of fresh vegetables from developing to developed countries is an even more profligate user of energy. The transport of asparagus, grown in Chile and flown to New York, has been estimated to consume more than 70 times the amount of energy as contained within the vegetable.[64]

The effect on the global climate from the transport of food is more than simply from greenhouse gas emissions, such as of carbon dioxide produced by the combustion of fossil fuels. Contrails left by jets and shipping wakes are thought to have distinct effects on the climate, separate to that of the associated greenhouse gas emissions.[65]

The Abuse of Ecosystems

Intensive farming practices have led to the transformation of ecosystems on a global scale. In most cases the clearing of forests, the planting of crops and the destruction of fisheries has enabled the feeding of more people. Free trade *per se* cannot be held responsible for all ecosystem damage. However, many ecosystems are harmed by modern trade in ways that are ignored by, or even deliberately disguised from, consumers. The law of supply and demand fails to reflect the real value of many ecosystem services and products.

Ground Water Contamination

In many places, fertilisation of the world's crops has led to substantial contamination of ground, river and coastal water with nitrogen, phosphorus and other substances. This nutrient loading has caused increased eutrophication, the symptoms of which pose a significant threat to coastal resources and ecological and human health.[66] These problems include the phenomenon of coastal 'deadzones', regions near river mouths in which algal blooms are depleted, not only reducing the spawning of fish. Large deadzones occur regularly in the Gulf of Mexico, the Baltic and off the Chinese coast.[67] Chronic exposure to nitrates in drinking water has been linked to gastric cancer.[66]

Fisheries Depletion

Supporters of the free market often claim that price signals will drive substitution and innovation. In fact, there is often a non-linear relationship between supply and price. That is, until supplies are very short, the price can be deceptively low. For example, in the case of fish, many depleted fisheries compete with others that are still plentiful, providing virtually no signal to contribute to conservation. Clever marketing can also mask scarcity and promote depletion. The long-lived, but now scarce deep-water fish called the orange roughy was originally known as the rather less appetising 'slimehead'.[68] Three-quarters of the fish sold in the United States as 'red snapper' belong to other species, at least some of which are likely to be endangered.[69] Consumers may thus inadvertently contribute to the depletion of fish species.

Further, not all species are ecologically equal; some may be the last member of an ecological 'suite', *i.e.* functionally related species that perform critical roles in maintaining ecological integrity. For example, several species of fish specialise in eating plankton in coral reefs, providing a form of ecological insurance. That is, the complete loss of *one* such species will not necessarily have a major ecological effect. On the other hand, the loss of the last population of a functionally related group is likely to have a major ecological effect, but this is most unlikely to be reflected in any price signal provided by the market. Parrotfish, one of a small number of species that feed on dead corals, are sold in London.[70] Few parrot-fish eaters are likely to have any idea that their purchase could have a disproportionately adverse ecological effect upon coral reefs.

6 Reforming the Global Economy

Externalities

Many of these shortcomings of free trade theory, whether involving the abuse of human beings, animals, the broader environment or the invisible contamination of food, fall into an ostensibly arcane and unimportant economic concept called 'externalities'. Formally, externalities are costs that are recognised but not systematically measured. By definition, the cost of externalities is not incorporated into the price of goods and services. Thus, there is almost no *economic* incentive to reduce externalities, which must rely instead on legislation, agitation and persuasion. If sufficient, these may force the mainstream economy to take them into account, 'internalising' the externality. For example, in developed countries, persistent campaigning and regulation has slowly reduced hazardous working conditions, at the same time increasing production costs and consumer prices.

Many factors prevent the incorporation of existing externalities into the calculation of prices. The most important is that the costs of externalities are nearly always borne by individuals, objects, and ecosystems which have or are utilised by people with no economic power (including people yet unborn). Markets respond to *effective* demand,[71] that is, to actual price and legislative signals, rather than to ideas that are simply desirable or laudable. For example, over 800 million people are chronically deprived of an adequate caloric intake; they clearly need food, but cannot command it. Similarly, the individuals, animals and ecosystems who are most harmed by externalities lack not only the power to change the current economic system, but, in most cases, to even raise the alarm.

The Evolution of Externalities As we have seen, the causal pathway between the consequence and cause of an externality is long and contested. Causal 'proof' may not only be impossible, but is sometimes not even suspected. Many ecological externalities developed because of the genuinely held view that the human impact upon nature could be no more than trivial. In the 19th century the naturalist Thomas Huxley claimed that marine resources were inexhaustible.[72] If this were really the case, then damage to oceanic ecosystems could be ignored in the balance sheet of fishing and land-based practices that harm coastal ecosystems, including coral reefs. A contemporary example concerns the relationship between fossil fuel consumption and climate change. Until recently, many people considered that human actions could not change the global climate, or that such change would be inconsequential.

At the time of the development of free trade theory, social conditions were vastly different from those of today. Food was grown everywhere by labour-intensive techniques only recently recognized as 'organic'. Animals were raised in fields and barns, and only draft animals and humans (including children) worked in factories. Empires, colonialism, social Darwinism, rigid class structures and racism were dominant values in much of the world. Democracy, as we now know it, was embryonic. The suffering of humans, especially those from the lower classes or from races held to be 'inferior', was considered part of the natural order.

In summary, there was virtually no incentive for economists to consider the kinds of adverse effects enumerated above, whether for humans, animals or the wider environment. Slowly, starting mainly in developed countries, social conditions changed, leading to an increased recognition of all kinds of problems. Problems that were near, visible and obvious attracted attention. Occupational safety, child labour and slavery (within developed countries) became legitimate areas for struggle and debate in the 19th century, assisted by the efforts of many social reformers. Unlike today, most workers lived close to the communities which consumed their produce. Injured workers were harder to ignore, a process assisted by increased literacy, good record-keeping and a free press. Virtuous competition led to improved working conditions and less environmental harm, especially when damage was local, visible and remediable. The pressure in developed countries to improve working conditions, provide compensation for injuries, and to make allowances for retirement all contributed to increased costs. Thus, these externalities were internalised. Prices increased, profits diminished, but a trade-off occurred in which the community as a whole benefited.

By contrast, in many countries today, rudimentary or even no compensation is available, and the cost to the consumer is correspondingly reduced. Even in developed countries, some of the hard-won gains for labour, described above, are being eroded. Prices are depressed by low wages, comparatively few taxes and low overheads. Some countries, such as India, are still scarred by ancient social divisions which sanction open discrimination, unacceptable in many countries. China too has an enormous underclass of disenfranchised 'floating' workers, with few rights.[73,74] Globally, this economic underclass forms a bedrock of human services upon which the broader economy is erected.

Real National Wealth

National accounting methods, such as those used to calculate the gross national product (GNP), routinely ignore externalities. There are many critics of the GNP who argue that it gives a misleading indicator of progress.[75] The GNP measures the circulation of formal currencies, ignoring barter, subsistence and the black economy. These exclusions understate national wealth, but this is outweighed by the failure to measure environmental damage and human suffering. For example, if an injured worker sues for damages, or if contaminated food causes an epidemic of gastroenteritis, the GNP will rise, giving the false impression that welfare has *improved*. Similarly, the conversion of an old growth forest into woodchips or paper will add to the GNP, while the loss of the natural capital represented by the intact forest is ignored.

Although pressure has increased, mainstream economic theory continues to resist proposals to internalise the cost of these externalities. Prices that do so are more expensive. Measuring, monitoring and incorporating these costs into prices constitutes a significant transaction cost. No government is likely to legislate to fully internalise these costs, because it would increase the price structure within that economy and disadvantage consumers and exporters in comparison to economies that act conventionally. As a result, goods with prices adjusted to incorporate externalities are likely to be restricted by market forces to niche status, such as 'fair traded' products.

Nevertheless, a fuller accounting of the externalities involved in the free trade of food, measuring what can be called 'real national' (or 'inclusive') wealth,[76,77] may reveal that a higher market price to the food-consuming customer is not quite as costly as first appears. A fraction of consumers who baulk at paying a premium for the 'gold standard' of fairly traded or 'organic' goods may still pay extra for produce that is *relatively* low in pesticide residues or heavy metals, or is not reliant on forced labour. Others, especially in affluent societies, may be willing to pay more if the extra funds can be diverted to environmental protection, or if the goods are produced using a less ecologically damaging method.

Collective action and pressure on large companies may reduce the profitability of selling goods produced using the most egregious methods, such as slave-dependent cocoa. More genuine development assistance to such societies is also likely to stimulate the internal social forces that can restrict the most barbaric production practices.

Morality

These uncosted problems collectively subsidise the production of abundant food for those with the economic means to pay for it, but constitute a hidden and immoral system. However, morality receives little recognition in modern economic theory.[78] All societies have a degree of inequality, and throughout history, affluence has almost always been attained, in part, at the expense of another person, animal or other environmental element. Nevertheless, today's immorality is arguably greater in scope than ever before, even if the worst cases of past immorality, such as the forced shipping of slaves from Africa to the New World, exceeds the worst of today.

Morality is influenced by time and culture, but a sense of right and wrong remains fundamental to religion, ethics, social cohesion and law. The gradual decline in morality, evident in many aspects of modern society, is illustrated not only by the examples such as those discussed above, but also by the dishonest accounting of firms such as Enron and the enormous growth in domestic inequality in many societies.[79,80] Although contested, the immorality of Western society is a plausible cause of much anti-Western terrorism.[25] Few commentators would dispute that, in turn, terrorists are also behaving immorally. The lack of attention to ethics and morality within mainstream economics may one day be seen as a grave error.

There are also many moral hazards in the restriction of free trade. In particular, even after decades of attempts, free trade continues to be practised very asymmetrically, with countries who have a stronger position (the European Union, Japan and the US, in particular) flouting principles of free trade with regard to commodities such as sugar, cotton, grain and skimmed milk powder.[6]

Fair Trade

Critics characterise fair trade as a threat to free trade, motivated by the desire to protect jobs at home against increased competition from the Third World, rather than reflecting genuine concerns about either the environment or human and animal rights. Fair traders have been described as 'irrational moral fanatics, prepared to sacrifice global economic welfare and the needs of developing countries for trivial, elusive, or purely sentimental goals'.[81] Examination of this statement in fact reveals language and assertions that could probably be fairly classed as propaganda. For example, the phrase 'global economic welfare' could be restated as 'global economic dominance'. The number of undernourished people (of macro or micronutrients) currently exceeds that of the total global population a century ago. Global inequality, no matter how measured, is more extreme than in any single country. At least 30 million people are frankly enslaved.[17] Such ills have not occurred because of the whims of fair traders, nor even because of the indifference of an impartial market. Instead, exploitation and inequality are cornerstones of the modern economy. Fair trade attacks this inequality. To further the stated goal of 'global economic welfare' more rather than less fair trade is needed.

7 Conclusion

Surely it is possible to develop a method to generate a trading system in which inequality is reduced and in which the most egregious abuses to humans, ecosystems and the wider environment are at least alleviated. Some advocates of free trade claim that the real purpose of the World Trade Organisation is, in fact, to improve well-being, including real incomes and living standards, for both rich and poor. But people with value systems that give genuine weight to the physical and moral hazards discussed in this chapter are likely to arrive at other conclusions.

This chapter has argued that the scale of recent and current exploitation – of other species, other humans, and other generations – is unprecedented, and is likely to have adverse consequences unprecedented in scale, including for many affluent populations. Just as enlightened individuals and groups lobbied in the past to reduce local exploitation, in part to reduce local social tension, we also need to better study the links between global injustice and global terrorism, for reasons of self-interest, as well as altruism. We also need to better document and study the physical hazards of food contamination. Despite claims to the contrary, the extent of physical harm from these substances may be very large, particularly for genetically vulnerable subgroups. Better labelling of contaminants would increase consumer awareness and stimulate niche markets. While entirely 'organic' food is likely to remain out of reach for most of us, new farming and industrial methods should enable the growing of food using more ecologically sustainable methods, with reduced concentrations of pesticides and other food contaminants.[82]

Improved mechanisation and computerisation will enable the replacement of many of the most menial, repetitive and poorly paid tasks by automata. The emergence of a more compassionate, inclusive and less fearful global society would accord greater recognition to the rights of other people, animals and ecosystems. Finally, the article has argued that a fuller and more balanced accounting of externalities will identify financial benefits likely to accrue from the reduction of physical and moral hazards. If this is possible, then a considerable volume of economic wealth currently consumed non-productively (such as treating preventable disease and insecurity) could be liberated. Much of this additional wealth could then be redirected to the poor, reducing inequality but without reducing the absolute living standards of the well-off. While not revolutionary, progress is possible.

References

1. D. Ricardo, *On the Principles of Political Economy and Taxation*, 3rd edn, John Murray, London, UK, 1821.
2. R. Wright, *Non Zero. The Logic of Human Destiny*, Pantheon Books, New York, NY, 2000.
3. A. Cigno, F.C. Rosati, L. Guarcello, *World Development*, 2002, **30**, 1579–1589.
4. O. Mehmet, *Westernizing the Third World: the Eurocentricity of Economic Development Theories*, Routledge, London, New York, 1995; F. List, *The Natural System of Political Economy*, Frank Cass, London, UK, 1983, translated by W.O. Henderson, originally published as *Systéme naturel d'économie politique* 1837.

5. P.A. Samuelson, *J. Economic Perspect.,* 2004, **18**, 135–146.
6. Oxfam, *Rigged rules and double standards. Trade, globalization and the fight against poverty*, 2002. http://www.maketradefair.com/assets/english/report_english.pdf.
7. A.L. Wright, W. Wolford, *To Inherit the Earth: The Landless Movement and the Struggle for a New Brazil*, Food First/Institute for Food and Development Policy, Oakland, CA, 2003.
8. J. Rosenberg, *The Empire of Civil Society: A Critique of the Realist Theory of International Relations,* Verso, London, UK, 1994.
9. M. Shanahan, C. Thornton, S. Trent, J. Williams, *Smash & Grab: Conflict, Corruption and Human Rights Abuses in the Shrimp Farming Industry*, The Environmental Justice Foundation, 2003. http://www.ejfoundation.org/pdfs/smash_and_grab.pdf.
10. F.E. Vega, E. Rosenquist, W. Collins, *Nature (London)* 2003, **425**, 343.
11. A. Bainbridge, Green Left Weekly 2001, **459**, http://www.greenleft.org.au/back/2001/459/459p6.htm.
12. M-C. Renard, *J. Rural Studies*, 2003, **19**, 87–96.
13. J.M. Keynes, The international control of raw material prices in *The collected writings of John Maynard Keynes* Vol. XXVII, MacMillan, London, UK, 1980.
14. H. Carter, The Guardian, Manchester UK, May 7, 2004, http://www. guardian.co.uk/ Refugees_in_Britain/Story/0,2763,1211528,00.html.
15. M. Busse, *World Development,* 2002, **30**, 1921–1932.
16. H. Wu, C. Wakeman, *Bitter Winds. A Memoir of my Years in China's Gulag*, John Wiley and Sons, New York, NY, 1994.
17. K. Bales, *Disposable People: New Slavery in the Global Economy*, University of California Press, Berkeley, CA, 1999.
18. C.D. Butler, *Global Change Human Health,* 2000, **1**, 156–172.
19. A.K. Sen, *Development as Freedom*, Oxford University Press, Oxford, UK, 1999.
20. G. Perusek, *New Politics*, 2004, **9**, http://www.wpunj.edu/~newpol/issue36/Perusek36.htm.
21. Food and Agriculture Organisation, *The State of Food Insecurity in the World 2002,* Food and Agriculture Organisation of the United Nations, Rome, 2002.
22. I. Darnton-Hill, *Aust. NZ. J. Public Health*, 1999, **23**, 309–314.
23. S. Grantham-McGregor, *Lancet,* 2002, **359**, 111–114.
24. B. Hayden in *Foundations of Social Inequality,* T.D. Price, G.M. Feinman (eds), Plenum Publishing Corporation, New York, NY, 1995.
25. C.D. Butler in *In Search of Sustainability,* R.M. Douglas, J. Goldie, B. Furnass (eds), CSIRO, Melbourne, Australia, 2005, 33–48.
26. J Gray, *False Dawn. The Delusions of Global Capital,* Granta, London, UK, 1999.
27. Survival International, *Last isolated Indians south of the Amazon make contact* 2004, http://www.survival-international.org/ayoreo.htm.
28. Human Rights Watch, *Tainted Harvest: Child Labor and Obstacles to Organizing on Ecuador's Banana Plantations*, Human Rights Watch, 2004, http://www.hrw.org/reports/ 2002/ecuador/.

29. International Labor Rights Fund, *Chocolate and Child Slavery: Unfulfilled Promises of the Cocoa Industry*, 2004, http://www.laborrights.org/projects/childlab/cocoa_063004.htm.
30. Human Rights Watch, *Turning a Blind Eye: Hazardous Child Labor in El Salvador's Sugarcane Cultivation*, Human Rights Watch, 2004, http://hrw.org/reports/2004/elsalvador0604/.
31. Anonymous, *Tea Coffee*, 2002, **176**, http://www.teaandcoffee.net/ 0102/ special.htm.
32. US Agency For International Development, *Summary of Findings from the Child Labor Surveys in the Cocoa Sector of West Africa: Cameroon, Cote d'Ivoire, Ghana and Nigeria*, US Agency For International Development, 2002, http://www.usaid.gov/press/ releases/2002/child_labor_iita.pdf.
33. J. Forero, *New York Times*, New York, NY, July 13, 2002, A1.
34. Friends of the Earth *Greasy Palms–Palm Oil, the Environment and Big Business*, http://www.eldis.org/static/DOC14908.htm.
35. M. Scully, *Dominion: The Power of Man, the Suffering of Animals, and the Call to Mercy,* St. Martin's Griffin, New York, NY, 2003.
36. D. Magnus, *Science (Washington, DC),* 2004, **306**, 58–59.
37. C.P. van Schaik, M. Ancrenaz, G. Borgen, B. Galdikas, C.D. Knott, I. Singleton, A. Suzuki, S.S. Utami, M. Merrill, *Science (Washington, DC),* 2003, **299**, 102–105.
38. R. Young, *The Secret Life of Cows,* Farming Books and Videos Ltd, Preston, UK, 2004.
39. J. Masson, *The Pig Who Sang to the Moon: The Emotional World of Farm Animals,* Ballantine Books, New York, NY, 2003.
40. A. Scott, *Nature (London),* 1999, **402**, 348.
41. S. Falkow, D. Kennedy, *Science (Washington, DC),* 2001, **291**, 397.
42. S.B. Prusiner, *Science (Washington, DC),* 1997, **278**, 245–251.
43. C.A. Llewelyn, P.E. Hewitt, R.S.G. Knight, K. Amar, S. Cousens, J. Mackenzie, R.G. Will, *Lancet,* 2004, **363**, 417.
44. E. Rivera-Milla, C.A.O. Stuermer, E. Málaga-Trillo, *Trends Genetics*, 2003, **19**, 72–75.
45. R.H. Devlin, M. D'Andrade, M. Uh, C.A. Biagi, *Proc. Natl. Acad. Sci. USA*, 2004, **101**, 9303–9308.
46. V. Shiva in *The Guardian Weekly*, 2001, April 12, 21.
47. D. Sharp, *Lancet,* 2001, **357**, 738.
48. D. Cyranoski, *Nature (London)*, 2004, **430**, 955.
49. J.A. Patz, P. Daszak, G.M. Tabor, A.A. Aguirre, M. Pearl, J. Epstein, N.D. Wolfe, A.M. Kilpatrick, J. Foufopoulos, D. Molyneux, D.J. Bradley, Members of the Working Group on Land Use Change and Disease, *Environ. Health Perspect.*, 2004, **112**, 1092–1098.
50. J. Kaiser, J. *Science (Washington, DC),* 2000, **289**, 371–372.
51. M. Eddleston, L. Karalliedde, N. Buckley, R. Fernando, G. Hutchinson, G. Isbister, F. Konradsen, D. Murray, J.C. Piola, N. Senanayake, R. Sheriff, S. Singh, S.B. Siwach, L. Smit, *Lancet*, 2002, **360**, 1163–1167.
52. D.J. Ecobichon, *Toxicology,* 2001, **160**, 27–33.
53. A.W. Kigotho, *Lancet,* 1997, **350**, 1528.

54. B. Ames, *Mutat. Res.*, 2000, **447**, 3–13.
55. C.D. Butler, A.J. McMichael in *Social Injustice and Public Health*, V. Sidel, B. Levy (eds), Oxford University Press, Oxford, UK, in press.
56. A. De Roos, S. Zahm, K. Cantor, D. Weisenburger, F. Holmes, L. Burmeister, A. Blair, *Occupat. Environ. Med.*, 2003, **60**, E11.
57. A. Menegon, P.G. Board, A.C. Blackburn, G.D. Mellick, D.G.L. Couteur, *Lancet*, 1998, **352**, 1344–1346.
58. J. Wargo, *Our Children's Toxic Legacy. How Science and Law Fail to Protect Us From Pesticides*, Yale University Press, New Haven, NJ, 1996.
59. A.P. Høyer, P. Grandjean, T. Jørgensen, J.W. Brock, H.B. Hartvig, *Lancet*, 1998, **352**, 1816–1820.
60. D.L. Davis, H.L Bradlow, *Sci. Am.*, 1995, **273**, 144–149.
61. R.A. Hites, J.A. Foran, D.O. Carpenter, M.C. Hamilton, B.A. Knuth, S.J. Schwager, *Science (Washington, DC)*, 2004, **303**, 226–229.
62. M. Durham in *The Guardian Weekly*, Aug 6, 2004, 18.
63. E. von Weizsäcker, A.B. Lovins, L.H. Lovins, *Factor 4. Doubling Wealth – Halving Resource Use. A New Report to the Club of Rome,* Earthscan, London, UK, 1997.
64. E. Millstone and T. Lang, *The Atlas of Food,* Earthscan, London, UK, 2003.
65. J.H. Seinfeld, *Nature (London),* 1998, **391**, 837–838.
66. R.E. Criss, M.L. Davisson, *Environ. Health Perspect.,* 2004, **112**, A 536.
67. A.S. Moffatt, *Science (Washington, DC),* 1998, **279**, 988–989.
68. D. Pauly, J. Alder, E. Bennett, V. Christensen, P. Tyedmers, R. Watson, *Science (Washington, DC),* 2003, **302**, 1359–1361.
69. P.B Marko, S.C. Lee, A.M. Rice, J.M. Gramling, T.M. Fitzhenry, J.S. McAlister, G.R. Harper, A.L. Moran, *Nature (London),* 2004, **430**, 309–310.
70. D.R. Bellwood, T.P. Hughes, C. Folke, M. Nyström, *Nature (London),* 2004, **429**, 827–833.
71. A.K. Sen, *Poverty and Famines: An Essay on Entitlement and Deprivation,* Clarendon Press, Oxford, UK, 1981.
72. H.S. Gordon, *J. Political Economy,* 1954, **62**, 124–142.; T.H. Huxley, *Nature (London)*, 1881, **23**, 607–613.
73. Y. Zhu, *Int. J. Population Geography,* 2003, **9**, 485–502.
74. A. Chan, *Human Rights Quarterly*, 1998, **20**, 886–904.
75. K.P. Arrow, P. Dasgupta, L. Goulder, G. Daily, P. Ehrlich, G. Heal, S. Levin, K.-G. Mäler, S. Schneider, D. Starrett, B. Walker, *J. Economic Perspect.*, 2004, **18**, 147–172.
76. C.D. Butler, *Lancet,* 1994, **343**, 582–584.
77. P. Dasgupta, *New Statesman*, 2003, **16 (781)**, 29.
78. R. Eckersley, *Well and Good: How We Feel and Why It Matters,* Text Publishing, Melbourne, Australia, 2004.
79. P. Krugman, *New York Times*, New York, NY, Oct 20, 2002, Section 6, 62.
80. J. Stiglitz, *The Roaring Nineties,* Penguin Books, London, UK, 2003.
81. R. Howse, M.J. Trebilcock, *Int. Rev. Law Economics*, 1996, **16**, 61–79.
82. G. Conway, G. Toenniessen, *Science (Washington, DC),* 2003, **299**, 1187–1188.

Subject Index